Bulbs

Essential know-how and expert advice for gardening success

CONTENTS

Bulbs

WHY GROW BULBS?

Bulbs bloom in such an array of magnificent forms and colours that there's virtually no limit to the displays you can create, in any type or size of garden. There are bulbs in bloom every month of the year, offering decorative interest at otherwise dull times, and lengthening the seasons by bridging flowering gaps. Bulbs are also great value, and most are incredibly easy to grow. They don't need much space or time, but offer spectacular results, whether used on their own as an eye-catching focal point, or as support to other plants in the garden.

FLOWERS ALL YEAR

Bulbs are essential plants for gardeners, because they offer colourful blooms in every season, from early spring all the way through to late winter. By planting bulbs, you can extend the period of interest in your garden so that it bursts with colour, shape, and texture for as long as possible, and has real visual appeal at what would otherwise be gloomy times of year.

Pineapple lily blooms are magnets for pollinators such as copper butterflies.

THE SHOW WILL GO ON

Many bulbs flower when there's little else going on in the garden. Some, for example, blaze happily away through autumn, long after most other flowers have already faded. Bulbs are packed with blooms that are guaranteed to brighten up dreary days – especially if it's cold and dark outside – and they bloom on, despite wind, rain, frost, or even snow. They draw us outdoors in winter and early spring with their joyful explosion of colours, adding sparkle to a stroll or creating a welcome view out of the window.

White and mauve blazing star and red montbretia light up late-summer gardens.

BLOOMS EVERY DAY

With a little planning, you can fill your garden with flowering bulbs throughout the year, from snowdrops and daffodils in February and March, through fritillaries and irises in April, May, and June, to July and August's agapanthus and gladioli. Late-flowering bulbs such as blazing star, dahlias, nerines, and montbretia light up September to November, and you can even bring colour and scent indoors in December and January by growing amaryllis and paperwhites (see p.42).

Early-blooming bulbs such as crocuses and winter aconites bring colour and cheer to the winter garden.

Purple alliums, cream irises, and white foxtail lilies carry the show in this gorgeous late spring to early summer scheme.

BRIDGING THE GAP

Bulbs are essential in dealing with awkward flowering gaps. Late tulips and alliums, for example, flower in a sort of mid-season lull period, after many popular spring perennials have finished their big display, but before a bounty of summer blooms appears. Flowering bulbs act as a bridge between the seasons, carrying the whole show through from one peak to another with tremendous ease.

You can also use them to help plug the empty spaces in planting schemes where other plants won't fit or work, because of their small spread and their ability to disappear for part of the year, allowing other plants room to shine. In this way, you can use your bulbs to fill almost every possible inch of earth with magnificent flowers.

CREATING A BUZZ

As well as being terrific for our own happiness and sense of wellbeing, having bulbs in flower throughout the year, especially in the off-season and between flowering peaks, is of huge benefit to pollinators. Insects such as bees, butterflies, hoverflies, moths, and beetles need to feed on flowering plants that are packed with nectar in order to survive. Species that emerge in the late winter and early spring, including the early bumblebee, and those that continue to buzz and flutter about late into autumn, are especially in need of blooms, and planting bulbs is the perfect way to support these vital creatures.

> **TOP TIP** PLANT YOUR BULBS TOGETHER IN GROUPS SO BEES AND BUTTERFLIES FIND THEM MORE EASILY, AND CAN ENJOY A NECTAR BANQUET IN ONE SPOT.

Flowering bulbs such as blazing star are a welcome food source for garden bees.

VERSATILE STYLE

Bulbs come in a vast range of forms, sizes, and colours, and you can use this diversity to continually change up your garden and create different schemes. There are tall bulbs and tiny bulbs, with blooms big and small, both bold and dainty, upright and nodding, and in every conceivable hue.

Whether you grow them on their own as an eye-catching focal point or use them to enhance and complement other plants, bulbs offer an easy way to design exciting displays in a wealth of styles and palettes. The only limit to the possibilities is your own imagination and creativity.

NEAT OR NATURAL

Many bulbs marry well with the highly ordered look of a formal garden, designed in a symmetrical layout with, for example, clipped hedges and evergreen topiary. Some bulbs will offer architectural foliage and geometric flower shapes as well as bright splashes of colour, which can act as a powerful contrast to the controlled structure of the neat and tidy green backdrop.

These blooms can also be a natural fit with informal gardens, such as rustic cottage plots. You can grow some bulbs in a loose, random fashion, so they jostle for attention among perennials and shrubs or drift en masse and self-seed and spread wherever they wish.

A calming white garden is here brought to life with tulips.

COLOUR CHAMELEON

Evoke different moods in your garden by using flowering bulbs in an array of colours. Make a soothing space by harmonizing pastel hues, such as pink hyacinths and purple crocuses. Stimulate the eye with complementary colours: orange 'Ballerina' tulips with blue grape hyacinth, for example. Clash bright colours, perhaps red gladioli with yellow begonia, or create a sense of calm using white and cream bulbs, including 'Thalia' daffodils and white camassias. Go for a relaxed but luxurious feel with jewel-coloured blooms like amethyst-purple alliums, sapphire-blue irises, and ruby-red tulips or peonies. Alternatively, pick sizzling-hot shades for a vibrant mix of orange dahlias, scarlet montbretia, and purple-leaved canna lilies.

Tulips combine beautifully with clipped topiary in a formal garden.

Alliums catch the eye in a relaxed and rustic cottage garden.

Even small gardens can evoke a woodland feel with the use of bulbs.

Striking, golden-yellow foxtail lilies pop in this prairie planting scheme.

SWITCH IT UP

The way in which most bulbs grow – putting on a big show with foliage and flowers over a few weeks, and then retreating quietly underground – means they can be used to easily change and refresh your planting colour schemes and styles during the growing season.

Create a container display (see p.34) with one group of bulbs flowering together, and then replace those with different bulbs the next month. Refresh your borders in this way too, by swapping blooms such as tulips for dahlias, growing them in bulb baskets or pots that can easily be popped in and out of the ground. Growing successively and using the same space for different blooms maximizes your opportunities to change colour palette and style so your garden is always fresh.

GET THE LOOK

You can use bulbs to achieve whatever planting style or look you want. Create a mini woodland idyll (see p.38) with dog's tooth violets, barrenworts, and wake robin. Follow planting ideas for prairie schemes using camassias and blazing stars. Make a sunny, Mediterranean-style retreat with African lilies. The diversity of bulbs makes them the go-to plants to complete your dream scheme, be it a meadow amassed with daffodils, or an urban courtyard with pots of calla lilies.

Swap containers of flowering bulbs in and out with ease.

Masses of daffodils and windflowers combine together to make a truly magical meadow.

BLOOMS FOR EVERY GARDEN

You may want to grow bulbs in pots, or fill gaps in a flower border, or create a rolling lawn spangled with blooms. Whatever your vision, every garden, big or small, can be enriched by bulbs. There are bulbs that will thrive in almost any growing environment, from sunny to shady spots and dry or damp soils, with plenty of options for every budget and level of experience.

Bulbs such as dwarf African lilies will even thrive in window boxes.

PERFECT FIT

Growing bulbs means you can enjoy the biggest flowering impact for the smallest amount of space. These are plants that generally have upright, narrow, linear, or strappy leaves that won't spread or take over. Many of them don't need much depth of soil, so there's a wealth of options available for even the tiniest of window boxes. Whatever the size of your growing patch – whether a windowsill, balcony, a small city courtyard, suburban back garden, or a country plot – you can and should grow bulbs. Compared to many other plants, such as perennials and shrubs, they don't take up much room.

Grape hyacinths occupy a small amount of space but have a tremendous impact in the garden.

THE PRICE IS RIGHT

Bulbs are great value, offering exceptional bloom for your buck, whatever your budget. They're often sold in bulk, in large nets or big bags, so you can buy lots of bulbs for the same price as just one potted plant and fill your garden with flowers for a lot less than ordering ready-grown blooms. Many bulbs are really easy to propagate and will even multiply on their own, making you many more plants completely free (see pp.24–27).

Low-cost bulk bags of bulbs mean more bulbs for less outlay.

ANY GARDEN

Each garden is different, but whatever your situation or soil, there's a bulb for you. For example, if you have an exposed coastal site, buffeted by wind and salt spray, then try red hot pokers. If your garden is in a city, where plants have to manage pollution and heat scorch, bearded irises are a good choice. Maybe you live on a steep slope, or garden in a frost pocket. Your plot could be south- or north-facing, sunny or shady. You might have soil that's dry, damp, sandy or clay, rich and fertile, or poor and free-draining. Often there's a mix of situations and soils within one garden. Whatever the advantages and challenges of your plot, you can find bulbs that will thrive there.

Bulbs are a terrific way of getting kids into gardening and involving them in the process of growing and caring for plants.

EASY FOR EVERYONE

Bulbs are incredibly easy to grow and you don't need specialist knowledge or experience to succeed with them. Unlike many other plants, most bulbs also don't need constant vigilance – you just pop them in the soil, walk away, and then rejoice when they pop up to bloom a few months later.

Simple and low-maintenance, bulbs offer spectacular results for a very small investment of time, effort, and money (see above). These are the ideal plants for beginners and busy people who still want to enjoy a gorgeous garden. They're also the perfect plants to grow with kids, and a lovely way to get the whole family interested and involved in gardening together.

Day lilies will grow in almost every soil and situation, whatever the challenges.

CHOOSING AND PLANTING

Bulbs are perfect little packages of flower power that come complete with everything they need to put on a terrific show. They range from true bulbs to corms, rhizomes, and tubers; there are hardy and tender types, and perennials and annuals, and they flower at various times. However, the basics are the same, as most garden bulbs don't need specialist knowledge or complicated techniques to grow. Just pop them in the soil and be patient – the results will be well worth the wait. In this section, you'll discover how to plant and grow bulbs, make more plants for free, and solve common problems.

WHAT IS A BULB?

A "bulb" refers to any swollen, knobbly, or fleshy root structure that has the potential to develop into a plant. Gardeners use this word to describe a broad range of plants that store their food and energy in an underground organ for when they're most needed. These plants gather nutrients in their bulbs during the growing season, then draw on the reserves to flower the following year.

Daffodils take in energy from sunlight while in growth, then store it in bulbs.

TYPES OF BULBS

When gardeners talk about bulbs, they're often referring not just to "true bulbs", but also to corms, rhizomes, and tubers. These are all geophytes (plants that store food and nutrients in underground structures), but they do have key differences.

True bulbs are made up of modified stems and leaves. The stem is compressed into a flat section at the bottom known as the basal plate, while the leaves are adapted to be fleshy and hold reserves of food and nutrients. Tunicated true bulbs, such as daffodils and tulips, are made up of tight layers of these fleshy leaves around a central bud, with a dry, brown, papery, outer layer or "tunic" wrapped around the outside. Scaly bulbs, such as lilies and some fritillaries, have more loosely packed layers, called scales, that overlap each other, and no tunic. They tend to be more easily damaged.

Corms, such as crocuses and gladioli, are adapted, thickened stem bases that appear solid inside, rather than layered like bulbs. Rhizomes, such as bearded irises, are underground stems that spread horizontally; every piece of a rhizome that has a bud on it will develop into a fresh plant. Tubers, for example dahlias, are also modified stems or swollen roots, which have produced growth buds or "eyes".

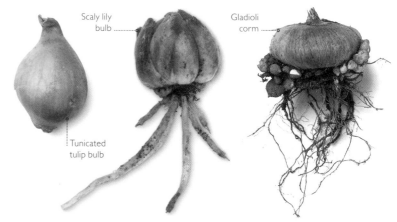

Scaly lily bulb ....

Gladioli corm ....

Tunicated tulip bulb

True bulbs can be tunicated (*left*) or scaly (*right*). Both are made up of many layers or scales.

Corms such as gladioli are solid, rather than layered like bulbs, and develop cormels on the basal plate.

Dahlia tuber

Iris rhizome

Rhizomes, such as irises, increase horizontally, with buds and shoots on the ends and roots underneath.

Tubers, such as dahlias, are rounded and fleshy, like fat fingers, with buds and shoots appearing in between.

Bulbs may look dead during their dormant period, but they're simply resting until the following season.

CHOOSING QUALITY BULBS

Bulbs are widely available at planting time from garden centres, specialist suppliers online, and many shops. When buying, choose big bulbs that are firm and plump – don't select damaged or soft ones. Squeeze them to see if they feel healthy. Avoid mouldy bulbs, which could have rot, and shrivelled or light ones, which may have dried out. If buying online, check the bulbs when they arrive. If any aren't up to standard, ask your supplier for replacements.

Check bulb quality before buying – they should look healthy and feel firm.

BULB DORMANCY

There's a huge variety of bulbs available, in all shapes, sizes, and colours, and with different growing needs. The one thing most bulbs have in common, however, is that they stop growing – or become "dormant" – for part of the year. This quiet period is brought on by a change of conditions, such as the turn of a season, and is totally normal.

After deciduous types have flowered, for example, their leaves yellow, turn brown, and wither. It may seem as though the whole plant has died or disappeared. However, this pause is part of the seasonal cycle – the bulb is still alive and healthy underground.

Dormancy is an ingenious and effective way for these plants to survive through difficult conditions, such as when it's very cold, or very hot. The nutrients and energy the bulbs have reserved see them through hard times and help them burst back into growth the following year. Having all their food stored up for when it's needed also means they often bloom at unusual times, when there isn't sufficient light or water available for other plants to flower.

SELECTING SIZES

Bulbs increase in size every year that they're grown, and will only flower once they reach a mature age and size: it can take several years for them to get large enough to store the amount of energy they require to bloom. This means that, in general, the bigger the bulb at the time you buy it, the more likely it is to produce flowers. Inevitably, this is reflected in the selling price, with large bulbs costing more than smaller bulbs of the same cultivar. The largest are sold as "top size".

For flower bulbs to "force" or grow indoors for early blooms, such as amaryllis, source the largest bulbs you can find.

Flower bulbs are measured by their circumference – the distance around the broadest section – and are usually described in centimetres. You may see tulips, for example, listed with their size given as "12/13", which means 12–13cm (4¾–5in) in circumference – "top size" in the case of these plants.

To ensure the most stunning displays of plants in your garden, buy top-size bulbs for those flowers that may only bloom once, such as tulips. Larger bulbs are also best for forcing flowers such as amaryllis and hyacinths. If you're willing to wait a little longer for results, it might be worth buying smaller, cheaper bulbs for plants that are reliably perennial and will increase or naturalize over time, such as daffodils.

PLANTING BULBS

Planting bulbs is easy – essentially, you just make a hole, pop them in, and then cover them with soil – but timing is everything. Spring-flowering bulbs are planted in autumn, summer-flowering bulbs are planted in spring, and autumn-flowering bulbs are planted in summer. Have patience, and you'll be amply rewarded when those small, dry, seemingly inert little bulbs burst into glorious bloom.

Dig your hole three times as deep as the height of the bulb.

You can start half-hardy or tender plants, like dahlias, into growth early under cover in spring by potting up in compost, ready to transplant outside later.

> **TOP TIP** IF YOU FIND A FORGOTTEN BAG OF BULBS IN THE SHED AND THE CORRECT TIME FOR PLANTING THEM HAS PASSED, DON'T THROW THEM AWAY. INSTEAD, PLANT THEM AS SOON AS YOU CAN AND SEE WHAT HAPPENS. THEY MIGHT NOT BLOOM WELL IN THEIR FIRST YEAR, BUT IT'S FAR BETTER THAN WASTING THEM.

WHEN TO PLANT

Early autumn is the best time to plant crocuses, daffodils, hyacinths, and other spring bulbs. Hardy late spring and summer bulbs such as alliums and lilies can be planted in mid-autumn, but tulips should be left until late autumn to protect them from viruses and pests. Tender summer bulbs such as gladioli and dahlias can be planted out once the risk of frost has passed in spring.

There are several bulbs that should be planted "in the green" – not when dry and dormant, but while they're in active growth. These bulbs – including winter aconites, snowdrops, and bluebells – are sold with leaves and sometimes even flowers on in early spring, having just been dug up. They should be planted immediately, to about the same depth as they were in the ground, which is easily identifiable as the point where the stems and leaves turn from white to green.

Some bulbs, such as snowdrops, have the best chance of success if lifted from the ground in active growth and transplanted "in the green".

PREPARATION AND DEPTH

Before planting your bulbs, prepare the ground. Dig it over and remove weeds. Most bulbs need a well-drained soil and will rot in wet or waterlogged conditions. You can improve your soil by adding organic matter such as well-rotted manure, garden compost, or leaf mould to the planting area. For better drainage, add grit. This will help with fertility and soil structure, but it's worth remembering that it's always best and easiest to grow the bulbs that are most suited to your garden's particular soil, climate, and conditions.

It's also a good idea to mix some slow-release fertilizer granules into the soil at the bottom of the planting hole.

Always plant at the correct depth. This varies according to the bulb, but a good guide is to plant three times as deep as its height. If in doubt, plant deeper – bulbs planted too shallowly may not flower or could have small blooms and short stems. The exceptions are some tubers and rhizomes, such as begonias and bearded irises – place these at or just under the soil's surface.

To plant a bulb, dig a hole wide and deep enough for the bulb, and place it with the pointed end facing up. Fill the hole with soil, firm gently, and water if necessary. If you're planting several bulbs, leave spaces of at least twice the bulb's width between them.

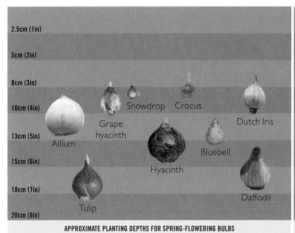

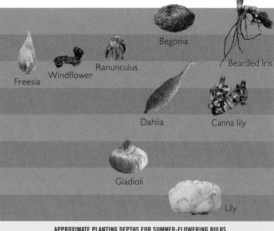

These planting depths are for some of the most popular spring- and summer-flowering bulbs.

CHOOSING YOUR TOOLS

Bulbs are often planted with a hand tool such as a trowel, but specialist tools called bulb planters can help with the task. They remove a plug of soil in one go, so you can put the bulb in the hole. Pushing it into the ground to create the next hole pops out the soil plug: you can use this to fill over the top of the first bulb. Some have helpful markings on them to show depth. There are also long-handled versions with foot plates.

If planting lots of bulbs, you could use a soil auger attachment with a cordless drill: simply drill into the soil once or twice, lifting out and dropping the loose soil beside the hole, pop in the bulb, and then replace the soil.

Tools such as this hand bulb-planter make the task of planting multiple bulbs much easier.

GROWING BULBS

After planting, most bulbs won't need any extra care or attention to start growing, so you can sit back and relax until you spot those fresh green shoots emerging through the soil. Once they're in active growth, you can help to keep your bulbs happy in a few easy ways, including making sure they're watered, well fed, and protected from the cold.

Bulbs don't need much attention until their leaves start to appear.

If the ground is already moist from a good shower of rain, there's no immediate need to water your bulbs after planting.

WATERING

Water your bulbs in well when you plant them if the soil isn't already moist from rain. After this, you can leave most of them alone, and not worry about irrigation until they begin to produce foliage. At this stage, if there's rain, they should be fine, but if the weather is dry and warm, it's best to water once a week or whenever seems necessary. Wait until the soil is beginning to dry out, as overwatering causes bulbs to rot. To check, press your finger gently into the soil to see if it's moist under the top layer. The soil in containers dries out quicker, so check them regularly and water them more.

You can reduce the frequency of watering when the foliage starts to die back, and stop once the leaves have gone brown and withered completely. Most bulbs prefer to stay dry during their dormant period.

If there's been no rain, drench the soil above the bulbs after planting.

FEEDING

If you wish to keep bulbs blooming brilliantly from year to year in the garden, you can give them a helping hand by boosting their nutrients while they're growing. There are two ways to do this. One method is to feed with a potassium-rich fertilizer, such as liquid tomato feed, once the leaves emerge from the ground. Dilute the feed to half the strength recommended on the pack.

In their first year after planting, limit feeding to once when the foliage appears, again after flowering, and possibly a third time, before the leaves die back. Once established, feed every two weeks during the growing season.

The second method is to wait until the bulb has finished flowering, and then begin using a general-purpose fertilizer regularly, following the instructions on the pack, for about six weeks, or until the leaves have begun to die back.

Bulbs will benefit from feeding with liquid fertilizer.

> **TOP TIP** DON'T ADD FERTILIZER THAT'S HIGH IN NITROGEN TO BULBS GROWING IN THE LAWN – IT WILL FEED THE HUNGRY GRASS MORE THAN YOUR BULBS.

Deadhead any spent flowers to keep your bulbs strong.

AFTER FLOWERING

Once blooms of bulbs are over, most can be deadheaded or the flower stems cut back. This stops energy being put into setting seed, sending it instead to the bulb underground, but if you'd like your bulbs to self-seed, leave them be. It's also not practical to deadhead individual plants in large-scale plantings.

Foliage is a different matter. The leaves of bulbs can become unsightly after flowering, often going yellow and leggy, looking ugly in mixed plantings. However, they must be allowed to

Never cut, tie, plait, or tidy foliage in the manner shown here.

grow on unhindered after flowering has finished because the bulb feeds itself through the sunlight that shines on its leaves. The plant soaks up the energy of the sun and photosynthesizes, or transforms it into nutrients and energy that it stores in its bulb. If you cut back, mow, tie, or knot the leaves, you'll stop the plant from being able to feed and gain the energy it needs to flower again next year. If the yellow foliage bothers you, grow other plants around it to help hide it as it dies back. Pots of bulbs can be moved to a less conspicuous area after flowering.

REPEAT BLOOMING

Some bulbs are perennial, like grape hyacinths and alliums, and come back every year to repeat their flowering display, so you can plant them once and just leave them in the ground. Others, such as many tulips, aren't as reliable, so it's best to plant new bulbs each year, as those left in the soil often get weaker each year or disappear. A select few plants have to be lifted out of the soil and stored while dormant (see p.23).

Those that stay in the soil all year will benefit from mulching with organic material, such as well-rotted garden compost or leaf mould. As the leaves die back, add a layer of about 5cm (2in) in and around the bulbs, being careful not to smother them. The mulch stops them drying out, keeps weeds down, and protects them from winter cold.

Protect perennial bulbs with a mulch of leaf mould.

LIFTING AND STORING

Some bulbs originate from warmer, drier climates than our own, and aren't equipped to survive cooler growing conditions year-round. Many of these bulbs are half-hardy or tender, and may not survive the cold of a harsh winter, while others need to remain **dry while dormant. To make sure they stay healthy and keep coming back into bloom, store them somewhere frost-free and dry under cover during their off-season. You can then replant them in the garden to flower the following year.**

Tender tubers like dahlias can be taken out of the soil and kept frost-free in winter.

Canna lilies can be grown in pots and brought inside during the coldest months.

PLANT PICKS

To ensure they'll survive and perform again next season, consider storing the following bulbs under cover during their dormant period:

Sword lily (*Acidanthera*) • Poppy anemone (*Anemone coronaria*) • Begonia (*Begonia*) • Coral drops (*Bessera*) • Canna lily (*Canna*) • Dahlia (*Dahlia*) • Pineapple lily (*Eucomis*) • Freesia (*Freesia*) • Summer hyacinth (*Galtonia*) • Gladioli (*Gladiolus*) • Corn lily (*Ixia*) • Star of Bethlehem (*Ornithogalum*) • Wood sorrel (*Oxalis*) • Tuberose (*Polianthes tuberosa*) • Persian buttercup (*Ranunculus*) • Harlequin flower (*Sparaxis*) • Bugle lily (*Watsonia*)

PROTECTING PLANTS

There are several popular frost-sensitive but perennial bulbs that gardeners traditionally protect over winter by storing them under cover. Plants such as canna lilies and glory lilies that are usually grown in pots are simply cut back and moved inside in their pots. Others, such as gladioli and dahlias, are lifted out of the soil completely before being stored.

Deciding whether you need to lift bulbs or not is a personal choice, and depends on your climate and the specific growing conditions of your garden. If you live in a warmer region or have a sheltered garden that remains mostly frost-free, you could take the risk of leaving these plants in the ground over the winter, with a protective mulch, to see if they'll come through.

However, if you regularly experience freezing temperatures, it's probably best to take them up and keep them inside through the colder months. Gardeners in damp climates or with heavy or waterlogged soil will also benefit from lifting these bulbs, which are prone to rotting in wet conditions.

Begonias like 'Yellow Giant' must be stored under cover while dormant.

LIFTING BULBS

Bulbs and corms such as the gladioli shown here can be lifted at the first sign of frost, when the leaves have turned brown and died back.

1 Dig up the bulbs or corms and carefully brush off the soil. Check for damage or disease and discard unhealthy ones.

2 Dry them by laying them out for a few days to a couple of weeks on a drying tray or wire cooling rack. Cut back the stems.

3 In the case of gladioli and other corms, separate the new corms from the withered old ones at the base. Discard the old corms, and dust the new ones and other lifted bulbs with fungicide.

4 Place bulbs and new corms in a net bag, a paper bag, or a box with scrunched-up newspaper. Hang or store them in a cool, dry, frost-free and well-ventilated place. Inspect the stored bulbs and corms periodically for rot or disease and dispose of any that look mouldy.

TOP TIP IF YOU'RE LIFTING AND STORING BULBS OF SEVERAL DIFFERENT CULTIVARS, MAKE SURE TO LABEL EACH BAG OR BOX SO THAT YOU KNOW WHICH IS WHICH.

STORING TUBERS

If storing plants such as dahlias and tuberous begonias, first cut back all the foliage and the flower stems to around 10cm (4in) above the ground. Dig up the tubers carefully with a garden fork, making sure not to damage them. Brush off most of the attached soil with your hands, but don't wash the tubers as you want to keep them as dry as possible.

Regularly check stored tubers and bulbs for health. It's also important to inspect them before replanting in spring.

Discard any tubers that look diseased or unhealthy or cut off sections that are damaged or diseased.

Sit or hang the tubers upside down somewhere cool to dry out for a couple of weeks, then bury them in a crate or box filled with dry compost or sand. Store in a dry, frost-free place, such as an unheated greenhouse, shed, or garage. Alternatively, store them on trays in a cool, dark space indoors, such as a cellar, with space for air to circulate between them. Check regularly for problems like mould, and discard any diseased, soft, or rotten tubers before they infect others.

EASY PLANTS FOR FREE

Bulbs are amazing plants that multiply all on their own. They naturally self-propagate, making mini copies of themselves, which you can then use to create more plants for free. The two principal methods of propagating bulbs are dividing them and cutting them up (see p.26). Division is an easy and straightforward process.

HOW BULBS REPRODUCE

The bulbs you plant in the garden reproduce by making copies of themselves underground as they grow. These are known as offsets, bulblets, cormlets, or side bulbs. The "baby" bulbs are attached to the basal plate, and are fed and nurtured by the original "parent" bulb for a few years, until they become large enough to flower themselves, and produce their own baby bulbs. Left alone, these clusters of multiplying bulbs will develop into clumps.

Gardeners can take advantage of this natural process by propagating these bulbs by division. This is a foolproof, risk-free way to increase your stocks and it produces results much sooner than growing from seed.

DIVIDING CLUMPS Lifting, dividing, and spreading out large clumps of bulbs is the easiest way to create sizable displays. Most bulbs should be divided after flowering. Bulbs planted "in the green" (see p.18), like snowdrops, should be lifted and divided while in active growth.

To lift and divide a clump of bulbs, such as daffodils, dig them up with a fork. Take the clump in both hands and gently pull it apart to create two or more smaller clumps. Replant these new clumps immediately, at the same depth as before, but spaced further apart.

Offsets such as bulblets can be easily removed from "parent" bulbs, like these *Iris reticulata*, once they're dug up.

GROWING ON OFFSETS

If you'd like to keep your large flowering bulbs putting all their energy into blooming, you can remove offsets such as bulblets and cormlets and grow them on yourself. Lift the plant after flowering, as the leaves are dying back, and pluck off the offsets from around the roots. Plant them at a depth of twice their own height in pots or trays of gritty, free-draining compost. After two or three years, the offsets will have grown large enough to flower. This method is great for corms and bulbs like crocuses and alliums. Some daffodils and tulips will produce side bulbs large enough to remove and replant in the ground right away.

> **TOP TIP** AS WELL AS BEING THE EASIEST METHOD OF PROPAGATION, DIVISION IS ALSO NECESSARY TO EASE CONGESTION. UNDIVIDED CLUMPS CAN GET SO CROWDED, WITH BULBS COMPETING FOR NUTRIENTS, THAT THEY STOP FLOWERING, SO MOST BULBS WILL BENEFIT FROM BEING DIVIDED EVERY FEW YEARS.

Dividing bulbs is the quickest, simplest, and best-value way to make many more plants for very little effort and investment.

UNEXPECTED TREASURES

Lilies produce offsets in several different ways. As well as the small bulbs that form at the base of the main bulb, many lilies also produce bulblets on the portion of the plant's stem that lies underground. These can be removed in autumn once the foliage has turned brown. Lift the bulb, separate the offsets, and pot them up to grow on.

Several lilies, including many Asiatic hybrids, also have aerial bulbils – offsets that develop on the plant above ground, where the leaves meet the stem. Small,

round, and usually brown or purple-black, these bulbils can be collected in late summer. Plant them in pots or trays, water, and leave somewhere frost-free. Seedlings will appear in a few weeks. They can be planted in the ground the following season, but may take several years to reach flowering maturity. Some alliums, such as *Allium sphaerocephalon*, also produce aerial bulbils, which can be propagated in the same way.

Aerial lily bulbils look like large, black seeds and can be picked out of the leaf axils where the foliage meets the stem.

SCALING

You can propagate scaly bulbs such as lilies and some fritillaries (*see p.16*) by "scaling" – using the outer scales of a bulb to encourage offsets to develop:

1 Once the leaves have turned brown, in late summer or early autumn, dig up and clean off the bulb. Snap off the outer scales, making sure to remove them as close to the base, or basal plate, of the bulb as possible. Replant the main bulb immediately.

2 Place a few handfuls of compost and grit or vermiculite into a clear polythene bag; add a little water to moisten. Dust the scales with fungicide and also place in the bag.

3 Blow into the bag and then knot or seal it to trap air. Store the bag in a dark, warm place such as an airing cupboard or a heated propagator, or in a warm room in a black plastic bag.

4 Check the bag every two weeks. After around four to eight weeks, but possibly up to twelve, you'll see small bulblets have formed on the scales. Pot up each scale individually in free-draining compost, with the bulblets just under the surface. Water, and grow on somewhere cool. You can plant them out the following autumn, but they can take up to five years to bloom.

MORE BULBS FROM ONE BULB

There are several simple methods for propagating bulbous plants other than by division, offsets, and scaling (*see pp.24–25*). Using a sharp knife in techniques such as chipping and scoring, you can significantly increase stocks of your favourite bulbs. Cutting methods can also be used to propagate tubers (especially dahlias) and rhizomes (including bearded and other irises).

Remove the old skin and roots and chop off the pointed top, or "nose", of the cleaned bulb using a sharp knife.

SCORING

Scoring with a knife is a simple technique that you can use to propagate true bulbs, such as snowdrops and hyacinths. It's also well-suited to increasing numbers of snake's head fritillaries.

When the bulbs are dormant, choose a good-sized, healthy specimen. Lift, clean, and dry it, and gently remove any withered or old dry bits of root or outer skin (tunic). Turn the bulb upside down. Using a sharp, sterilized knife, cut a straight, shallow groove into the basal plate at the bottom, and another at a right angle, to form a cross. Dust the cuts with fungicide, then put the bulb into a pot of moist sand, with the top facing down and the basal plate showing. Place the pot in a dark and warm spot, such as a heated propagator or airing cupboard, for about 10 weeks, until little bulblets appear from the cuts.

Replant the bulb upside down in a pot of compost, with the bulblets at the top, just under the surface. Water and leave in a frost-free place. The following autumn, take off the bulblets and plant them in their own pots. They should then flower in three to four years.

Scoring can be used to increase stocks of rare snowdrops like 'Blewbury Tart'.

Cut the bulb in half through the base, in half again, and then halve each quarter so you have eight "chips".

CHIPPING

Daffodils and alliums are among the bulbs that can be propagated by "chipping". Choose a healthy, dormant bulb. Clean, dry, and remove old skin and roots, then cut off the "nose". Turn the bulb upside down. Cut through the base, dividing into eight "chips", each with its own section of the basal plate, and dust with fungicide. Treat the chips the same as scales (see p.25): place them base-down in a clear polythene bag in moist vermiculite. Blow into the bag to fill it with air, then seal and label it. Store in a warm, dark place. Bulblets will appear on the chips in 12 weeks or so. Pot the chips individually in compost with the bulblets lightly covered. They should flower in three to four years.

SPLITTING TUBERS

You can propagate plants like dahlias and peonies by dividing their root tubers. Peonies are best divided in autumn. Dahlias can also be divided at this time, when you dig them up to store over winter (see p.23), but the optimum time is in spring, when you're preparing to pot up the stored tubers.

Discard any tubers that are damaged or mouldy. Place the tuber on your work surface. Look for the pinky-red growth eyes or buds, which appear around the bases of the short stems left on top.

Each division you make must have its own piece of stem with growth eyes. With a sharp knife, cut down cleanly through the centre of each stem, and separate out the resulting pieces. Dust the cut surfaces with fungicide and plant them individually in pots of compost. Only plant into the ground outside once there's no risk of frost. Dahlia divisions can flower in their first year. Peonies can take two or three years to bloom again.

Use a sharp knife to divide dahlia tubers, and ensure that each section contains a length of stem with buds.

SPLITTING RHIZOMES

Rhizomatous plants like bearded irises can be divided just after flowering.

1 Dig up a large clump of rhizomes using a garden fork. Remove the excess soil and pull apart large sections – using two forks back to back will help prise them apart.
2 Cut off the woody old section of the rhizome, leaving fresh pieces with young leaves and roots attached.
3 If splitting bearded irises, trim the leaves to 15–20cm (6–8in) from the base. Use sloping cuts to make a fan shape. Trimming helps minimize the risk of wind and water damage.
4 Replant the fresh pieces of rhizome immediately, with the top of each one just under the surface of the soil.

NEED TO KNOW
- Always choose healthy bulbs to propagate and take precautions to stop the spread of disease.
- Sterilize the knife with disinfectant between batches of bulbs.
- Apply sulphur powder or fungicide to the cut sections; you may choose to wear gloves to avoid skin contact with chemicals.
- Keep your work surface clean.

PESTS AND DISEASES

Bulbs are relatively trouble-free, but they can occasionally fall prey to pests and diseases. Some can be managed or cured with controls, but several require that you remove and destroy infected plants completely, to prevent spread. The best way to avoid these problems is to practise good hygiene when propagating and storing your bulbs.

LILY BEETLES

PROBLEM Patches of leaves and buds eaten. Black, slimy lumps appear on stems and leaves.
CAUSE The bright red lily beetle eats the plants and lays eggs; the black substance is the grubs' excrement.
REMEDY Pick off adults and grubs when seen. Spray with systemic insecticide only when plants are not in bloom.

EELWORMS

PROBLEM Distorted leaves and stems; late flowering. Browning to layers inside the bulb, which can be seen when cut open horizontally.
CAUSE Stem and bulb eelworm: a tiny nematode not visible (or barely visible) to the human eye.
REMEDY Remove and destroy affected bulbs. No chemical controls available.

GREY MOULD

PROBLEM Fuzzy, light grey fungal growth appears on foliage or flowers, particularly if they're damaged. Infected tissues turn brown, rot, and collapse.
CAUSE A fungal disease called *Botrytis*.
REMEDY Remove affected bulbs quickly and destroy them. Don't replant bulbs in the same spot. Avoid overcrowding plants and encourage good air circulation.

NARCISSUS BULB FLIES

PROBLEM No leaves or flowers, or very few; yellow, stunted growth. Maggoty grubs inside soft, infected bulbs.
CAUSE Larvae eating the bulb underground. In late spring, the fly lays eggs on the bulb; the larvae then hatch and burrow into the base to feed.
REMEDY Remove and destroy affected bulbs immediately.

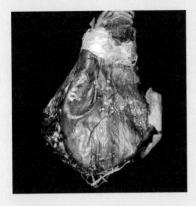

ROT

PROBLEM Fungus on bulbs in storage; soft, rotten bulbs. Yellow, stunted growth on planted bulbs and no flowers.
CAUSE Fungal infections, including fusarium basal rot, gladioli dry rot, and black slime disease.
REMEDY Bin or burn affected bulbs – don't add them to compost. Improve air circulation in storage.

SQUIRRELS AND RODENTS

PROBLEM Bulbs dug up after planting or damaged while they're in storage.
CAUSE Feeding of squirrels and other rodents such as mice and rats.
REMEDY Plant bulbs deep. Cover freshly planted pots or clumps with chicken wire; cover large plantings with netting. Home remedies include grating soap into the planting hole.

TULIP FIRE

PROBLEM Distorted shoots, leaves and stems; scorched, yellow or brown spots on foliage; mouldy flower buds and brown spots on bulbs.
CAUSE A fungal disease that's exclusive to tulips.
REMEDY Destroy affected plants immediately. Don't plant tulips in the same area for at least three years.

SLUGS AND SNAILS

PROBLEM Holes in leaves or foliage and flowers disappear.
CAUSE Feeding of slugs and snails, especially at night or in wet weather.
REMEDY Clear debris. Seek out and remove slugs and snails by hand. Cover soil with grit or eggshells. Try copper strips around pot rims, and beer traps and slug pellets on the ground.

THRIPS

PROBLEM Streaking on gladioli leaves, distorted flowers, and stunted growth.
CAUSE Small sap-sucking insects called thrips peel off outer layers of foliage.
REMEDY Prevent gladioli thrips by cutting down top growth on corms you've dug up for winter and drying the corms before storing. Destroy any plants that are already affected.

VIRUSES

PROBLEM No flowers, or weak ones; blooms with colour breaking, or flamed patterning; mottled or streaked foliage. Reduced fertility and bulb growth.
CAUSE Viruses, including tulip break virus and lily mottle virus.
REMEDY Destroy affected plants. Control sap-sucking insects such as aphids and thrips, which carry viruses.

WAYS TO GROW

Bulbs are immensely versatile plants that can be used to enrich almost every type of garden and a wide range of growing environments, from shady woodland schemes to dry gravel gardens. They're the backbone of the classic summer border and pack a punch in pots layered with a succession of blooms. Whatever your style, size of garden, or overall vision – perhaps a lawn spangled with delicate spring blooms or a house bursting with glorious scent and colour in winter – in this section you'll find inspiration and practical advice on how to make bulbs work for you.

MIXED BEDS AND BORDERS

Flowering bulbs will bring zing to mixed borders, pairing up beautifully with herbaceous perennials and shrubs, adding value by filling in gaps, and bringing interest early and late in the year as well as in the lulls between seasons. You can also use bulbs to create exciting seasonal bedding schemes on their own or with colourful annuals, laid out in a planned or random fashion.

Bulbs planted in baskets can be easily lifted in and out of borders.

Make an impact with a colourful, eye-catching bedding scheme of tulips, fritillaries, and daffodils.

Plant bulbs such as hyacinths with wallflowers in autumn in a narrow bed along a path.

ALL-BULB BEDS

For really stunning seasonal displays, choose a bed to use just for bulbs (or for bulbs and annual plants). The plants can then be lifted after flowering and replaced with those that bloom later, so you'll get two or more displays in the same space every year. These bedding schemes take a little planning, but mean that you can experience masses of colourful blooms all at once in a really sensational show.

Plant spring-flowering bulbs such as hyacinths, tulips, and fritillaries in autumn, either on their own or with annual and bedding plants that flower at the same time, such as forget-me-nots and wallflowers. While they're preparing to bloom, start off summer-flowering bulbs such as poppy anemones and Persian buttercups, or later-blooming gladioli, dahlias, and cannas, under cover in pots. Then you can replant the whole bed with them and annuals such as cosmos, once all risk of frost has passed.

MIXED BORDERS

Although many bulbs are recommended for planting in borders, some manage the competition for space, water, and nutrients in these situations better than others. Bearded irises and lilies, for example, will live in the ground year-round in borders without a problem, and should be given the same space as the herbaceous plants in the scheme.

Perennial bulbs, such as spring-flowering daffodils and grape hyacinths, can be planted around established perennials. Dig holes in between the plants carefully, to prevent damaging their roots, and pop in the bulbs. Place 15–30cm (6–12in) from the base of the plants – closer for smaller or later-growing perennials, and further away for vigorous or large perennials. This means that when you come to divide your border plants, you'll be less likely to dig up or damage the bulbs.

One-season bulbs such as tulips can be planted into bulb baskets and lifted out cleanly and easily as a whole group after flowering, reducing the disturbance to the border. Then annuals such as cosmos and salvias and later-blooming, tender bulbs such as gladioli or dahlias can be dropped into the same gap, to give a full and vibrant look to the border all summer.

Pink tulips bring a mixed border of herbaceous perennials such as spurge and annuals like forget-me-nots to life.

DESIGN DETAILS

For impact, use at least 25 to 50 bulbs in a scheme, planted in groups of six or more. Alternatively, plant them in bands or a ribbon through a bed or border. Always plant tall bulbs such as lilies towards the back of the border, and low-growing bulbs towards the front, so they'll be seen. Dotting them about individually results in a meagre, spread-out feel with no wow factor.

Create a truly integrated display by choosing just a few cultivars in complementary colours and planting them in large numbers, rather than a few each of many different types. You can achieve a pleasingly random distribution effect by pouring these chosen bulbs into a bucket together. Then take handfuls and gently scatter them across the spaces in the border. The resulting scheme will be cohesive and fluid rather than rigidly regimented.

DISGUISING FOLIAGE

The leaves of many spring-flowering bulbs will go yellow as they start to die back and go dormant, which looks messy in a border scheme. You can't remove the fading foliage of perennial bulbs until it has withered (*see p.21*), so the best way to deal with this issue is to use any surrounding herbaceous plants as camouflage. Plant bulbs like daffodils with late-spring, fast-growing plants such as peonies, hostas, hardy geraniums, lady's mantle, and ferns, which will produce their fresh foliage at the right time to disguise the yellow leaves. Once the bulbs have died back, the other plants will fill the space left by the dormant bulbs above ground. The worst offenders, such as alliums, whose leaves fade before they flower, can be planted behind leafy perennials like columbine and hardy geraniums.

> **TOP TIP** IT'S OFTEN HARD TO REMEMBER EXACTLY WHERE YOUR BULBS ARE ONCE THE FOLIAGE DIES BACK, LEAVING JUST BARE SOIL, SO IT'S A GOOD IDEA TO MARK THE SPOT WITH A SHORT BAMBOO CANE.

Foliage from poppies, columbines, and catmint fill in the gaps around a striking planting of alliums.

Leaves of *Euphorbia griffithii* conceal the fading foliage of grape hyacinths.

BULBS IN CONTAINERS

The most versatile way to grow bulbs is in containers, which allow you to create satisfying miniature schemes in one pot, to flower all at once or in sequence over the season. You can constantly update larger displays by swapping pots of bulbs in and out as they flower. Containers also allow gardeners to easily grow bulbs that might not be suitable for their garden's soil or climate.

Small daffodils such as *Narcissus* **'Tête-à-Tête'** are perfect for pots.

Plants bulbs like lilies at the same depth in containers as you would in the ground.

Top pots of bulbs with grit to keep them moist and clean.

BEST BULBS FOR OUTDOOR POTS

African lily (*Agapanthus*) • Windflower (*Anemone*) • Begonia (*Begonia*) • Canna lily (*Canna*) • Crocus (*Crocus*) • Pineapple lily (*Eucomis*) • Dwarf gladioli (*Gladiolus nanus*) • Hyacinth (*Hyacinthus orientalis*) • Dwarf iris (*Iris reticulata*) • Lily (*Lilium*) • Grape hyacinth (*Muscari*) • Small and species daffodils (*Narcissus*) • Striped squill (*Puschkinia*) • Squill (*Scilla*) • Tulip (*Tulipa*), particularly Fosteriana, Single Early, and Triumph types

Triumph tulips such as 'Couleur Cardinal' work well in containers.

PLANTING IN POTS

To plant up a container with bulbs, choose a pot with a drainage hole at the bottom. Cover the hole with "crocks" (pieces of broken pottery), a rock, or large pebble to stop soil from leaching out. As with planting in the ground, bulbs in pots should be placed at their recommended depth – around three times the height of the bulb. Place compost in the pot to the right depth, and pop in your bulbs.

You can generally plant more bulbs closer together in pots for the best display – just make sure they're not touching each other or the edges of the pot. Cover with compost, up to about 2cm (1in) below the rim of the pot. Top with a layer of grit or decorative gravel, which stops wet soil from splashing up on the leaves and flower petals during watering or from rain. Water well and, if necessary, cover the top of the pot with chicken wire to protect from snacking squirrels in autumn and winter.

Create a stunning display of bulbs by arranging different sizes and heights of containers together, each filled with one variety of bulb, such as tulips, daffodils, and camassias.

CONTAINER CARE

Bulbs growing in containers need to be watered more often than those in the ground, because the soil dries out quicker. Keep the soil just moist through winter, not damp, and start watering regularly once shoots appear. Keep going until the foliage begins to die back. Container-grown bulbs also need regular feeding throughout the growing season.

Bulbs chosen for a one-season show, such as non-perennial tulips, can be taken out of the pots after flowering and put on the compost heap, but hardy perennial bulbs, such as daffodils, crocuses, and grape hyacinths can be kept and regrown in the garden. You can transplant bulbs from pots into the ground after flowering or once they're dormant, and then use the pots for other bulbs and plants if you like.

However, many smaller bulbs, such as dwarf irises and miniature daffodils, are suitable for growing in pots continuously. Some, including snowdrops, need to be repotted each year into fresh compost.

Container bulbs may need extra protection in winter, as the soil in pots is more likely to freeze and damage them. Move pots to a sheltered spot or wrap the outside of them in fleece or bubble wrap if cold weather is forecast.

LAYER UP

A "bulb lasagne" is a clever way of planting different bulbs in layers, one on top of the other, so you get the most impact from flowers over the longest period, in the smallest space. This method means you can have several different types of bulb flowering at one time in a pot for a

Layer your container by planting larger bulbs towards the base, and smaller ones nearer the top.

beautiful mixed display, or plant for a succession of early, mid-, and late-season blooms in just one container.

To create the lasagne, place a layer of compost at least 10cm (4in) deep at the bottom of the pot, then put in the largest bulbs. Add another layer of compost and the next set of bulbs, between the points of the first set. You can have two or more layers, depending on pot size – but because of the density of planting, use fewer bulbs and space them further apart than if planting one type of bulb in one layer.

NATURALIZING IN GRASS

There are few more joyous sights in spring than a beautiful green sward dotted with vibrant flowers. Growing flowering bulbs such as crocuses, daffodils, and fritillaries en masse in this way, so that they multiply and spread as they would in the wild, is known as naturalizing. This is often done with woodland plants around trees in larger gardens, but it's also a great idea when you want to add colour and interest to any lawn, big or small, or decorate grassy slopes or banks that are difficult to mow.

Miniature narcissus and fritillary plants enliven this garden lawn.

THINKING AHEAD

Planting bulbs in an existing lawn demands a little planning, especially if you want to keep your sward short and tidy. You won't be able to mow until the foliage of your bulbous plants has died back, which is usually about six weeks after flowering, so the best tactic is to choose bulbs that flower early, before grass growth is at its peak.

If you're planting bulbs in a grassy area that you're happy to leave unmown for longer periods, you can create a meadow with a much broader choice of plants, and even have a succession of flowers blooming in the tall grass throughout the season: for example, early snowdrops and scillas followed by crocuses and daffodils, camassias, and alliums.

Always choose bulbs that are suited to your garden's conditions. For rapid establishment and spread, select species bulbs, and those known to self-seed, such as grape hyacinths.

PLANTING UP

To naturalize in grass, plant spring-flowering bulbs in the autumn and summer- and autumn-flowering bulbs in the spring. Mow the sward short and avoid planting bulbs at regular intervals, which will result in a regimented look. To achieve the artfully wild appearance of a natural meadow, take a handful of bulbs, scatter them in a sweeping motion over the lawn, and plant them where they fall.

To plant larger bulbs, such as daffodils, cut into the turf using a trowel or bulb planter and remove the plug of soil. Put the bulb in the hole, pointed end up, and twist it gently into the earth. Cut a second hole to push out the original plug of turf and sprinkle some soil from the plug over the bulb. Push the plug back flush with the lawn to cover the bulb.

For smaller bulbs, such as crocuses, you can plant numbers of bulbs closer together. Use a spade to cut two flaps of turf and lift them back to each side. Loosen the soil beneath with a fork and scatter the bulbs over the area. Score the underside of the turf and then roll it back in place, firming with your hands.

To multiply your stocks, lift and divide clumps after flowering every three years (see pp.23–24). Don't deadhead self-seeders like snowdrops and crocuses.

> **TOP TIP** FRITILLARIES AND SOME SMALLER DAFFODILS MAY BE DIFFICULT TO ESTABLISH IN GRASS. INSTEAD, GROW A FEW BULBS SEPARATELY IN POTS, THEN PLANT THEM INTO THE GRASS IN THEIR FINAL POSITIONS WHEN THEY REACH A HEIGHT OF ABOUT 5CM (2IN).

A bulb planter is a plug-cutting tool that makes naturalizing bulbs in lawn quick and convenient.

Smaller bulbs can be planted in groups by lifting areas of turf.

Randomly planting bulbs in grass can create the look of a natural meadow.

TOP BULBS FOR NATURALIZING IN GRASS

SHORT GRASS

Early crocus (*Crocus tommasinianus* and its cultivar 'Ruby Giant') • Snake's head fritillary (*Fritillaria meleagris*) • Common snowdrop (*Galanthus nivalis*) • English bluebell (*Hyacinthoides non-scripta*) • Daffodil (*Narcissus* 'February Gold' and 'Tête-à-Tête') • Glory of the snow (*Scilla forbesii*) and Siberian squill (*Scilla siberica*) • Wild tulip (*Tulipa sylvestris*)

LONG GRASS

Persian onion (*Allium hollandicum*), yellow garlic (*Allium moly*), and honey garlic (*Allium siculum*) • Camassia (*Camassia leichtlinii* subsp. *suksdorfii* Caerulea Group) • Byzantine gladioli (*Gladiolus communis* subsp. *byzantinus*) • Grape hyacinth (*Muscari armeniacum*) • Daffodil (*Narcissus* 'Actaea' and *N. poeticus* var. *recurvus*) • Drooping star of Bethlehem (*Ornithogalum nutans*)

Crocuses are ideal early-flowering bulbs for neat lawns.

COOL, DAMP, AND SHADY SPACES

Gardens with a lush, green forest feel are soothing to the senses, and there's nothing better than escaping into the cool shade during hot weather. Few plants will thrive where shade is deepest, but there are some bulbs that will tolerate full shade and many that will happily grow and flower with only a few hours of light a day.

Fumewort is a versatile, hardy plant that will grow in moist or dry shade and in pots.

Autumn crocuses will naturalize in grass in the part shade of the tree canopy, and look great under autumn foliage.

Wake robin is a top choice for difficult moist, shady areas.

UNDER TREES

It's challenging to create successful planting in shaded areas, especially under trees – they cast shadows and can outcompete other plants' roots for moisture, space, and nutrients in the soil. However, bulbs that originate from cool, moist, woodland habitats will appreciate the same conditions in the garden.

Very little will grow in the dense, dry ground under evergreens, but plenty can thrive around deciduous trees and shrubs, as they allow sunlight through

their leafless canopies when bulbs need it most, in winter, spring, and autumn. As a result, most woodland plantings and shade-tolerant bulbs peak in spring or autumn. In deciduous woodland, the soil is often moist from rain in spring and autumn, but dry in summer, just as these bulbs like it. The earth is also humus-rich from the layers of leaves that drop each year. You can emulate these conditions in any garden, with shade and mulch.

To plant right up to the base of the trunks of trees or shrubs, choose small perennial bulbs that will naturalize over

time, like cyclamen and lily of the valley, so you'll only have to disturb the tree's roots once. Using a narrow trowel, hand fork, or dibber, loosen the soil between the roots and carefully make holes large and deep enough to plant the bulbs. Water in and mulch with leaf mould.

Bulbs that require more moisture, such as wake robin, or a few hours of sunlight a day, such as bluebells, will do better further out under the canopy or on its fringes. They'll also do well under trees with a light canopy, such as birches, which cast dappled shade.

Arum lilies shine brightly among the ferns in this small, shaded town garden.

SHADY SPACES

It's not just woodland gardens that have to deal with shade – urban gardens that are shaded by buildings face the same challenges. The areas where structures cast their shadows are often permanently dull and can be plagued with damp and moss, especially in north- and east-facing gardens, which may also be colder.

In these areas, plant bulbs that are shade- and moisture-tolerant, such as fumewort, bleeding hearts, snowflakes, and grape hyacinths. Brighten up a dim or dark corner by using shade-tolerant bulbs with white or pale blooms, such as arum lilies, and those with variegated leaves. Plant pots of colourful bulbs that will grow in containers in part shade, such as martagon lilies, to make the most of every inch of available space.

BULBS FOR SHADE

LATE WINTER AND EARLY SPRING
Windflower (*Anemone blanda*) • Wood anemone (*Anemone nemerosa*) • Eastern cyclamen (*Cyclamen coum*) • Winter aconite (*Eranthis hyemalis*) • Snowdrop (*Galanthus*) • Daffodil (*Narcissus* 'W. P. Milner') • Striped squill (*Puschkinia*) • Glory of the snow (*Scilla luciliae*)

MID-SPRING AND EARLY SUMMER
Lily of the valley (*Convallaria*) • Barrenwort (*Epimedium*) • Dog's tooth violet (*Erythronium*) • Bluebell (*Hyacinthoides non-scripta*) • Bleeding heart (*Lamprocapnos*) • Snowflake (*Leucojum*) • Solomon's seal (*Polygonatum*) • Wake robin (*Trillium*)

MID- TO LATE SUMMER AND AUTUMN
Autumn crocus (*Colchicum*) • Martagon liles (*Lilium martagon*) • Lily turf (*Liriope*) • Toad lily (*Tricyrtis*)

MOISTURE-LOVING BULBS

Most bulbs rot in cool, damp, or wet conditions, but there are several that make great additions to gardens with poorly-drained soil or to bog-garden plantings, moist meadows, and around ponds. Arum lilies and flag irises, for example, are so water-friendly they can be planted into the shallow depths of ponds, either in baskets or as marginals at the edges. Hardy marsh orchids and snakeshead fritillaries prefer a moist soil in an open location, while snowdrops and dog's tooth violets can manage both shade and damp. Schizostylis and society garlic are two later-flowering bulbs that will also tolerate wetter soils, but they like their faces in the sun.

Snakeshead fritillaries and wood anemones like moist areas near a pond.

HOT, DRY, AND SUNNY SPACES

Gardens in areas of low rainfall or with poor, free-draining soil that doesn't hold moisture are extremely challenging sites to find suitable plants for. This is also true of gardens that are baked by the sun all summer long – for example, those in the countryside, exposed at the coast, or sheltered in the city. However, all is not lost: there are some bulbs that cope well with these difficult, hot and dry conditions.

Dwarf daffodils such as 'Hawera' and species tulips always work well in rock gardens.

ROCK GARDEN BULBS

Rock gardens or rockeries are usually used to grow small alpine plants that originate in mountainous areas, with sharp drainage and plentiful sunlight. These places are cold in winter; they're also dry, which is important for many hardy bulbs that hate winter wet. Rock and crevice gardens, scree beds, alpine troughs, and planted drystone walls usually peak in late winter and spring, making them perfect for some bulbs. These sites can have another flush of interest in autumn, when some bulbs bolster the late-season display.

TOP PICKS *Spring bulbs:* crocus • dwarf iris • false shamrock • fritillary • miniature daffodil • snowdrop • species tulip • squill • starflower • star of Bethlehem • windflower • *Autumn bulbs:* autumn crocus • autumn daffodil • autumn snowflake • ivy-leaved cyclamen

DROUGHT-TOLERANT SCHEMES

Due to the need to conserve water and adapt to our changing climate, plantings that require less irrigation are on the rise. They suit sites with light, sandy, or stony soils that drain freely, are of low fertility, and tend to be sunny and dry. Mulching the ground with a top dressing, such as gravel, helps to keep moisture in. Many bulbs work well in such schemes, which often look their best from late spring through summer.

TOP PICKS *Spring and early summer bulbs:* allium • bearded iris • crocus • Dutch iris • foxtail lily • grape hyacinth • Portuguese squill • species tulip • *Mid- to late summer and autumn bulbs:* colchicum • corn lily • freesia • nerine • Peruvian lily • pineapple lily • society garlic • sword lily

Peach-coloured foxtail lilies light up a gravel garden with blue grassnut flowers.

SEASIDE SELECTIONS

One of the most difficult places to grow plants is on the coast. Gardens by the sea are exposed and open to the elements, because of lack of trees and shrubs in these areas. They typically experience hot, baking sun and have very sandy, fast-draining soil that's low in nutrients. They also have to deal with strong and salt-laden winds, which destroy the lush green foliage and delicate flowers. However, even in these most challenging gardens, some bulbs will still put on a show.

TOP PICKS *Summer bulbs:* African lily • day lily • red hot poker • *Autumn bulbs:* bugle lily • montbretia • nerine

Red hot pokers can manage the salt, wind, and sun of a beachside garden.

PRAIRIE PLANTING

A terrific choice for areas that are open, hot, sunny, and somewhat dry, but have a richer, more fertile soil is a prairie-style planting scheme. This naturalistic look, inspired by North America's grasslands, emulates those dense plantings with garden schemes that mix perennial flowers with ornamental grasses, and bulbs that peak in late summer. However, the major contribution of bulbs to these plantings is in extending the season of interest by flowering before the main event, in spring and early summer, when the foliage of the prairie plants makes a fabulous, green foil to flowering bulbs.

TOP PICKS *Spring and early summer bulbs:* camassia • foxtail lily • Siberian flag iris • *Late summer bulbs:* blazing star • day lily • drumstick allium

Drifts of camassias will bring spring interest to prairie-style planting.

Drumstick alliums will thrive and multiply in sunny, open schemes.

INDOOR BULBS

When plants are grown to flower out of season or earlier than usual, it's called "forcing". This method is often used with bulbs to produce blooms indoors during the coldest months of the year, tricking the bulbs by mimicking the change from winter into spring.

Place the bulbs so they're just poking out of the surface of the soil.

Enclose in a black plastic bag to prevent any light from entering.

FORCING BULBS

To force most bulbs to flower, you need to imitate winter and then spring, by first creating a period of cold and dark before bringing them into warmer, brighter conditions. Although different bulbs require slightly different processes, the general method is as follows.

Choose a container with drainage holes that's at least twice as deep as the bulbs. For best results, your growing medium should be bulb fibre – a specialist, free-draining mix that's perfect for growing bulbs indoors and out. However, a light, loam-based compost will suffice.

Put grit or gravel in the base of the container for drainage. Add a layer of moist growing medium. Pop the bulbs in on top, close to but not touching each other or the edges of the pot.

Top up with growing medium, with the noses of the bulbs just poking out at the surface.

Most bulbs need a dark phase, so put the pot in a black plastic bag to block the light, if necessary, and keep in a cold place (such as a shed or cellar) for 10–15 weeks, at a temperature of between 1.5°C (35°F) and 10°C (50°F). Check every two weeks, and water if needed: the medium should be moist, not wet.

Once the fresh, often whitish, shoots reach about 3cm (1¼in), take the pot out of the bag and place it somewhere cool and light (such as a conservatory, porch, or cold frame) for two weeks to green up the leaves. Once green, place the pot somewhere warm and bright, above 15°C (59°F), to bloom. A windowsill is ideal, but avoid siting it close to a radiator or in direct sun. Keep well-watered while in growth.

BULBS FOR FORCING

Poppy anemone (*Anemone coronaria*): approximately 10–12 weeks from planting to flowering • Crocus (*Crocus*): 10–18 weeks • Amaryllis (*Hippeastrum*): 10 weeks • Hyacinth, prepared (*Hyacinthus orientalis*): 11–13 weeks; hyacinths, not prepared (*Hyacinthus orientalis*): 13–17 weeks • Early iris (*Iris reticulata*): 15–17 weeks • Grape hyacinth (*Muscari*): 16–18 weeks • Paperwhite daffodils, (*Narcissus*): 5–10 weeks; other daffodils (*Narcissus*): 17–19 weeks • Tulip (*Tulipa*) 15–20 weeks

Crocuses need a period of cold and dark to bloom.

Paperwhites like 'Ziva' are simpler than crocuses – just plant and wait.

EASY HYACINTHS

Hyacinth bulbs for sale in early autumn have often been pre-chilled or "prepared" for forcing. This means they have already experienced their period of cold and will flower much sooner, typically over the festive season. You can also buy ordinary, non-prepared hyacinth bulbs in early autumn and put them in the fridge in a paper bag for about 4–6 weeks to get them to the same point.

The easiest, mess-free way to have indoor blooms is to grow hyacinths in water, using forcing jars or vases – shaped glasses with a bowl or cup at the top. Fill the bottom with water to just below the base of the top section, and place your bulb in the cup. Alternatively, put some pebbles or glass beads in the bottom of any container or glass vase, add the bulbs on top, and water to the level just below the base of the bulb.

Forcing jars, such as these, are the most straightforward way to grow hyacinths indoors.

FLOWER, STORE, REPEAT

Hippeastrum bulbs, commonly known as amaryllis, can be stored after blooming to grow again the following year. Remove any spent flower spikes but leave the foliage to grow on. The pots can be placed out in the garden during the summer. In late summer or early autumn, stop watering and allow the foliage to dry out and die back. Cut down any yellow leaves or stems. Put the pots in a cool place such as a shed or garage for a couple of months. Then bring them back into the house, placing them somewhere warm and bright, and begin watering and feeding to start them growing again. Amaryllis and paperwhite daffodils such as 'Ziva' (see *image, opposite*) don't require cold or dark periods like other forced bulbs.

Hippeastrum 'Minerva' can be encouraged to bloom again every year.

Plant amaryllis in a pot that's slightly wider than the bulb.

GOOD FOR SCENT

Fragrant plants fill the air with seductive smells, so it's no wonder that bulbs with scented blooms are in such high demand, both as border beauties and as cut flowers. You can enjoy their appealing aromas to the fullest in the garden, and indoors, by choosing the best varieties, and planting and growing them in the right spot to make the most of their magnificent perfume.

Beneficial insects, including bees and butterflies, will flock to gardens that have scented flowers such as peonies.

FABULOUS FRAGRANCE

Scented flowers contribute yet another layer of richness to your garden, an additional sensory dimension beyond sight, touch, and sound. Our sense of smell is intricately linked with both our psychology and emotions – a fragrance can lift the spirits, and even just a brief waft of perfume from a certain bloom can evoke vivid memories from childhood, holidays, or special events.

You can enjoy the indulgent delights of scented bulbs throughout the entire year, from heady hyacinths indoors during winter and honey-scented snowdrops in early spring, through to the sweet-scented lily of the valley in springtime, and powerful, spicy-scented lilies and fruity-fresh freesias from summer to autumn.

Scented flowers also attract beneficial insects, including pollinators such as bees and butterflies, to your garden (see p.9).

TOP SCENTED BULBS

Lily of the valley (*Convallaria*) • Freesia (*Freesia*) • Some snowdrops (*Galanthus* 'Atkinsii', *G. elwesii*, and *G.* 'S. Arnott') • Sword lily (*Gladiolus murielae*) • Some day lilies (*Hemerocallis lilioasphodelus*) • Bluebell (*Hyacinthoides non-scripta*) • Hyacinth (*Hyacinthus orientalis*) • Bearded iris (*Iris germanica*) • Dwarf iris (*Iris reticulata*) • Lily (*Lilium*) • Grape hyacinth (*Muscari*) • Some daffodils (*Narcissus poeticus* cultivars, including 'Actaea', 'Cheerfulness', and any Jonquilla and Tazetta daffodils) • Some peonies (*Paeonia lactiflora* and *P. officinalis*) • Tuberose (*Polianthes tuberosa*) • Some tulips (*Tulipa saxatalis* and some cultivars, including 'Angélique' and 'Ballerina')

Some day lilies such as *Hemerocallis lilioasphodelus* have lovely scents.

'Ballerina' is one of a few tulips with a delightful fragrance.

SITES FOR SCENT

Make the most of your scented plants by growing them in situations that will help to intensify their magnificent fragrance. The perfume will be more highly concentrated, and also retained for a longer period, in warm, sheltered, and enclosed locations such as walled gardens, sunken gardens, and courtyards.

Plant scented bulbs where you'll receive most enjoyment from them: for example, by the front door, along a pathway, or near an open window, so you get wonderful whiffs every time you pass. Place small and lower-growing blooms at the front of borders so you don't miss out on their charms, or plant them in raised beds or pots on an outdoor table, closer to nose level.

Create space for scented bulbs around your patio or seating areas, especially for those, such as tuberose, that will continue to release their perfume on summer evenings.

Place pots of hyacinths on your patio or along a path in spring so you can enjoy their fragrance when you pass by.

INDOOR AROMAS

You can enjoy scented bulb blooms in the house as well as the garden, either by forcing and growing them in pots (see pp.42–43) or by cutting the flowers for posies and bouquets.

Cut your blooms in the morning. Use secateurs to snip them off around 2cm (¾in) from the bottom of the stem at a 45-degree angle, then place in water. Tepid to warm water is usually best for cut flowers from the garden, but cut flowers from spring bulbs need cold water. Leave them in water for a few hours, in a cool, dark place, before arranging. Strip off any foliage that would sit under water in the vase. Fill a vase with water and add half a teaspoon of bleach per litre (quart) of water.

To keep cut flowers fragrant and fresh for as long as possible, display them away from direct sun, draughts, heat sources, such as radiators, and bowls of fruit. Take out fading flowers and change the bleach-infused water every few days.

A few fragrant flowering bulbs need a little extra thought when cutting. Don't mix daffodils in a vase with other blooms, as they secrete an inhibiting substance. Cut peonies when their buds are still closed but firm and fat; leave double-flowered types until just before they open. When cutting lilies, leave at least a third of the stem on the plant, and remove the pollen-laden anthers from the stamens to prevent staining interiors.

Narcissus **'Actaea'** has a wonderful perfume, both in the garden and the vase.

Snowdrops, crocuses, and winter aconites are a glorious addition to any garden. They kick off spring in spectacular fashion with their bright and cheery blooms – the earliest of the season.

WINTER AND EARLY SPRING

The growing season for bulbs begins with the coldest months of the year, when a host of beautiful plants bring colour and life to beds, borders, and containers. Swathes of delicate, white snowdrops start the season, before early irises and gorgeous deep-blue grape muscari take over and carry the show into spring. These hardy little plants are such a welcome sight, toughing out the elements at the harshest time of year and holding firm through frosty days. They send up green foliage from the frozen ground before producing heart-warming blooms that light up even the darkest days of winter and early spring.

CROCUS *CROCUS*

These much-loved dwarf beauties have goblet-shaped blooms in shades of purple, blue, orange, yellow, and white. At their centres stand bright orange stamens ripe with pollen – a real treat for early bees and pollinators. With only a little care, crocuses will come back year after year and multiply easily.

BULB TYPE Corm
HARDY Fully hardy
HEIGHT 8–15cm (3–6in)
LEAF Grass-like, striped; deciduous
POSITION Full sun, part shade

'Ruby Giant' is an early blooming crocus with large flower cups. It spreads well and can be planted under trees.

CHOOSE

Spring-flowering crocuses are among the most popular garden bulbs. They have a broad variety of flower colours and flowering times, from late winter through spring. Most crocuses have dark green leaves with a central pale stripe, which helps to identify and differentiate them from other spring bulbs when the flowers aren't out.

Species crocus are small and dainty, and bloom early in the season. One of the first to flower is *Crocus tommasinianus*, early crocus, which is a great choice for naturalizing. The cultivar 'Ruby Giant' is bigger

and blooms a deep, rich purple with a bright orange stamen. It's more tolerant of shade than other types.

C. sieberi, Sieber's crocus, is another early type. 'Tricolor' is a pretty cultivar with lilac flowers, a white ring at the base of the petals, and a yellow centre.

C. chrysanthus, snow crocus, flowers profusely and is often bi-coloured. 'Gipsy Girl' has golden cup-shaped blooms with flame-like brown to purple markings flaring up the outside of the petals. 'Prins Claus' is white, brushed with bold dabs of dark purple.

C. vernus (Dutch crocus) hybrids, such as the pure white 'Jeanne d'Arc' and purple-veined 'Pickwick', bloom

later. The hybrids are tougher than the species types, with larger corms and flowers, and cope better when planted in rough or dense grass.

PLANT AND CARE

Crocuses open their flowers in the sunshine, but close them up in shade, or when it's overcast or raining. They give their most striking display in an open position in full sun, but are also happy in partial shade under deciduous trees and shrubs as long as they receive several hours of sunlight each day in spring. Avoid planting your crocuses in dense shade under evergreens or at the base of north-facing walls.

Plant spring-flowering crocuses in gritty or very well-drained, poor to moderately fertile soil. They need

A fragrant variety, 'Tricolor' has pretty, pointed petals in three colours.

'Gipsy Girl', with its bronze-striped blooms, looks glorious even when closed.

White 'Jeanne d'Arc', like most Dutch crocus hybrids, flowers later in the season.

moisture in autumn and spring, but the ground must be free-draining as they'll rot in soggy conditions.

Plant the corms with the pointed end facing up, at a depth of 8–10cm (3–4in), spaced at least 5cm (2in) apart. They look best in small groups of six to ten, or en masse to create a glorious carpet of flowers. If grown in pots, add grit to the compost to ensure good drainage.

Water crocuses well after planting and in spring if conditions are dry, but not in summer when the plants are dormant.

After flowering, allow the foliage to turn yellow and wither before removing, or simply let it die back

completely. If you're growing crocuses in a lawn, don't mow the grass until after the leaves have died back.

Squirrels and mice sometimes dig up and eat the freshly planted corms, but you can deter them by covering large areas such as sections of grass with netting after planting. For smaller areas – a section of a border, for example – bury a piece of chicken wire just under the soil where the corms are planted. You can cover pots with wire mesh too, bending it over the top edges. Rodents seem to really like Dutch crocus corms, but tend to leave early crocus alone.

PROPAGATE Crocuses will self-seed and spread happily on their own. However, large clumps can get congested and stop flowering, so dig up and divide these groups in autumn – you can then separate them into smaller clumps and replant them further apart. At the same time, remove any little cormlets and pot them up in compost to grow on and plant out the following autumn.

> **TOP TIP** IF YOU'RE PLANNING TO PLANT CROCUSES EN MASSE FOR A DRAMATIC DISPLAY, LOOK OUT FOR "LANDSCAPE PACKS" OF BULBS CONTAINING 100 OR 500 CORMS – THESE ARE FAR BETTER VALUE THAN BUYING SEVERAL SMALL PACKS.

WAYS TO GROW

Crocuses look wonderful in large drifts and are excellent for naturalizing under trees or in grass. They're also a good choice for brightening up the front of borders, rock gardens, and containers. Some varieties of Sieber's crocus and snow crocus have fragrant flowers: plant them in small pots, bulb bowls, or window boxes and at a height (on a table or window sill, for example) where their delightful scent will be easier to catch as you pass by.

Plant crocuses in big swathes for a spectacular, high-impact show.

EASTERN CYCLAMEN

CYCLAMEN COUM

BULB TYPE Tuber
HARDY Fully hardy
HEIGHT 8–10cm (3–4in)
LEAF Rounded, patterned, or mottled; deciduous. Spreads to 10cm (4in)
POSITION Part shade

Eastern cyclamen flower over a long period – from winter through early spring – and, once established, are extremely long-lived. They produce unusual white, pink, or red flowers with reflexed petals, like shuttlecocks. The leaves are equally interesting, with an array of silvery patterns and markings.

Pewter Group cyclamen have eye-catching, silver-sheened leaves.

CHOOSE

These flowering perennials look small and sweet in the winter garden but are exceptionally tough and hardy. The foliage appears in late autumn, followed by flowers. These can be white, pink, or carmine, with some varieties having a contrasting colour at the petal's base.

The most widely available cultivars are sold simply by their flower colours, such as white and shades of magenta and rose pink. The Pewter Group has pink blooms and foliage that looks as though it's been washed in silver. 'Maurice Dryden' has white flowers and silver-centred leaves with a green edge.

PLANT AND CARE

Choose a sheltered spot for your tubers, protected from cold winds and hot sun – around the base of deciduous trees or shrubs is ideal, as long as the soil isn't too dry. Plant them in any well-drained, moderately fertile, humus-rich soil in autumn at a depth of 5cm (2in). The roots should be facing down and the growing point facing up. If you can't see roots, plant them with the concave side of the tuber facing down.

Cyclamen growing in the ground will benefit from an annual mulch after flowering. For pot-grown plants, add grit and leaf mould to the compost at planting time and water just before they dry out. Soak the base of pot-grown plants well when needed, rather than regularly sprinkling water over the tops of flowers. Feed pot-grown plants with a slow-release fertilizer in spring.

WAYS TO GROW

Don't crowd Eastern cyclamen with lots of other types of plants – grow them where they'll have good air circulation and be best appreciated, such as in a rock garden, in pots, and on slopes and banks. They're a good choice for tricky, shaded corners and at the bases of shrubs and trees.

Cyclamen tubers may be eaten by mice and squirrels. In damp conditions, the plants are prone to attack from grey mould (see p.28), so plant them in a suitable location with good air flow and drainage. Removing spent flowerheads can help to reduce the risk of attack.

PROPAGATE Eastern cyclamen self-seed and spread naturally. Gently clearing the area around them of weeds and deep leaf build-up after flowering will help the dropping seeds to reach bare soil.

If you wish to collect seed, wait until the flower stem coils up after blooming. Harvest the ripe seed and soak in water overnight. Sow on the surface of a 50:50 mix of seed compost and grit, then cover with a thin layer of sieved compost. Keep warm in a heated propagator, greenhouse, or clear plastic bag out of direct sunlight until seedlings appear.

Fresh seed can be harvested to grow once the flower stem has coiled up.

WINTER ACONITE

ERANTHIS HYEMALIS

With their tiny butter-yellow blooms, winter aconites bring some welcome cheer to the new year. They hug the ground, producing a low whorl of stem leaves, like a collar or ruff, around eye-catching, chalice-shaped flowers. Perennial and wonderfully fuss-free, they're also great for early pollinators.

BULB TYPE Tuber
HARDY Fully hardy
HEIGHT Up to 10cm (4in)
LEAF Divided; deciduous. Spreads to 10cm (4in)
POSITION Part shade
WARNING! Ingestion may cause stomach upset; contact with leaves may irritate the skin. Wear gloves when handling.

Winter aconites have golden, goblet-shaped flowers circled by a green ruff.

CHOOSE

Winter aconites are wonderful groundcover plants that create a carpet of early colour. They're great for shady areas under trees where grass won't grow – they cover up the bare ground in winter with flowers, and then during spring with foliage as their basal leaves appear. These plants are very hardy and, given the right growing conditions, come back reliably every year.

Eranthis hyemalis, the species winter aconite, is easy to source and, once established, will spread with ease. 'Flore Pleno' is a more showy cultivar with double flowers, while 'Schwefelglanz' has pale apricot to cream-coloured blooms. If you want to feature them as pot-grown plants – rather than to self-seed and naturalize – try the Turbergenii Group sterile hybrid 'Guinea Gold', with its golden blooms and bronze-green leaves. For the biggest cup-shaped flowers and fine filigreed foliage, look out for the *E. hyemalis* cultivar 'Orange Glow'.

PLANT AND CARE

Like many woodland plants that flower early in the season, winter aconites prefer a spot that mimics being under a canopy. Here they'll receive sunshine in spring, before the deciduous trees leaf out and offer summer shade. Choose a partially shaded spot in moist but well-drained, humus-rich soil. Improve your soil with rotted manure or garden compost before planting, if necessary.

Dry tubers are available in autumn. Soak them overnight before planting. However, you'll get the best results by using tubers in the green in spring – they should be in leaf, freshly dug, and planted immediately on arrival. Plant tubers 5cm (2in) deep, and spaced 5cm (2in) apart, if planting in small groups.

WAYS TO GROW

For a colourful winter display, plant under early-flowering shrubs such as witch hazel and Japanese quince, or around the bare legs of rose bushes. They're a great choice for gardeners who have deer and rabbit problems. Winter aconites are shallow-rooted, meaning they'll also grow well in pots and window boxes.

Winter aconites need to stay moist throughout their dormant season during the summer months. Water them when conditions are dry, especially if the plants are still getting established. Apply a mulch of leaf mould or garden compost to plants after flowering. Allow the foliage to die back naturally.

These low-maintenance plants are generally not affected by pests but they can suffer from a fungal disease called eranthis smut, which causes the stems to split open with black spores. Dispose of infected plants immediately.

PROPAGATE Winter aconites can sometimes be tricky to establish and don't like to be disturbed. To propagate, it's best to allow them to self-seed. Over time, when left to themselves, they can form large drifts.

SNOWDROP *GALANTHUS*

Snowdrops herald the end of winter with their pure-white blooms and fresh green foliage. The flowers have three outer petals and three smaller inner ones with green markings on their tips. Planted en masse, snowdrops light up the garden. Admired up close, they offer a gloriously subtle honey scent.

BULB TYPE Bulb
HARDY Fully hardy
HEIGHT 12–30cm (4¾–12in)
LEAF Linear, strap-shaped; deciduous
POSITION Part shade
WARNING! Ingestion may cause stomach upset; contact may irritate the skin. Wear gloves when handling.

CHOOSE

The several different species and many hundreds of cultivars of snowdrops all look quite similar, so it's sometimes tricky to distinguish one from the other. You can, however, tell them apart by the varied green or sometimes yellow markings on the inner petals; the outer petals may also be tipped with colour.

There are both single and double varieties of snowdrop and these vary in both size and flowering time. Many types are rare or scarce, and demand high prices per bulb, but these are best left to specialist collectors. You can easily create wonderful displays with more widely available, affordable types.

Galanthus woronowii, giant snowdrop, is the species most commonly found for sale. It grows to around 15cm (6in) tall and has notably bright and fresh green leaves. *G. nivalis*, common snowdrop, reaches the same height, but has slimmer, grey-green leaves. The cultivar 'Flore Pleno' is the double-flowered version and its intricately layered inner petals resemble a dancer's tutu. Like most double plants, it doesn't produce seed so won't spread by self-seeding.

G. elwesii, greater snowdrop, has larger flowers than the common snowdrop and blooms earlier. It reaches up to 30cm (12in) high, has broader, glaucous leaves, and is more tolerant of sun than the others.

The double-flowered 'Flore Pleno' has a large, complex ruffle of inner petals.

G. plicatus, pleated snowdrop, is a prolific flowerer that's fuss-free, easy to grow, and reaches 18cm (7in) high. *G. plicatus* 'Wendy's Gold' and *G. nivalis* Sandersii Group are unusual for their yellow rather than green markings.

G. 'S. Arnott' is a hybrid type that blooms between the early and latest snowdrops. Reliable and vigorous, it doesn't set seed, but it's a great choice for increasing your plants through division as it will soon make big clumps in the garden. *G.* 'Atkinsii' is another hybrid that spreads well in this way; it also flowers early.

> **TOP TIP** WILD SNOWDROP PLANTS ARE OFTEN DUG UP ILLEGALLY AND SOLD AS CULTIVATED PLANTS, SO SOURCE YOUR BULBS FROM A REPUTABLE SUPPLIER.

Great for naturalizing under trees, *Galanthus elwesii* is one of the tallest and earliest snowdrops and also has some of the biggest blooms.

'Wendy's Gold' is one of the few snowdrops with yellow details.

PLANT AND CARE

Snowdrops are perennial and, once established, will be low-maintenance and relatively trouble-free. However, to really thrive, they'll need the right growing conditions from the outset. Unlike other spring flowers, snowdrops will flower for longer when it's cool and overcast. Their blooming time is shorter in bright, warm conditions.

Plant snowdrops in a moist but well-drained soil that won't dry out during the summer. Often, they'll fail or not come back after one season due to free-draining soil or too much sun. You'll get better results and longer displays by placing them in part shade, where they'll get some sunlight and rain in spring but cover in summer – under the canopy of deciduous trees is ideal.

Bulbs can be planted in autumn, 10cm (4in) deep, with the pointed end facing up. Alternatively, you can plant them "in the green" (see p.18), in late spring, which often results in them establishing more easily. It's always best to transplant them as soon as they arrive so they don't dry out. Plant them to the same depth as

'S. Arnott' has exquisite, v-shaped green markings on its inside tips.

they were growing before – marked by where the stem turns from green to white. Space them 10cm (4in) apart.

If you're growing snowdrops in containers, be sure to repot them into fresh compost every year when dormant, in late spring or summer. Plants in containers must be watered to stay moist throughout the year.

Divide your snowdrops while they're still in active growth, or "in the green".

After flowering, allow the foliage to die back on its own. Snowdrops can be affected by slugs and narcissus bulb fly, and occasionally by grey mould (see pp.28–29). Squirrels are attracted to dry bulbs planted in autumn.

PROPAGATE To increase your stocks, lift clumps of snowdrops after flowering, and before foliage turns yellow. Divide them into smaller groups and replant straight away at the same depth, further apart. For types that set seed, harvest the fresh green seed as it ripens and sow into pots of compost. Sown plants may not match the original plant. Other propagation methods include scaling (see p.25) and chipping (see p.26).

Display snowdrops on a shelf or stand to admire their nodding blooms.

WAYS TO GROW

Snowdrops look fantastic in large drifts and they naturalize well. To best admire them up close, display different types higher up on a bank. Bring them closer to eye-level in pots on a plant theatre or stepped stand. This way, the varied green markings inside the hanging flowers can be more easily seen.

AMARYLLIS *HIPPEASTRUM*

Amaryllis are "forced" or grown indoors for their large, showy flowers, which offer a bold and colourful show in winter and early spring. The flamboyant trumpet- or streamer-shaped blooms top tall, sturdy stems. With the right care, these gorgeous bulbs will reflower every year.

BULB TYPE Bulb
HARDINESS Tender
HEIGHT 25–90cm (10–36in)
LEAF Fleshy, strap-shaped; deciduous
POSITION Indoors
WARNING! Ingestion may cause stomach upset. Wear gloves and wash hands after contact.

Amaryllis bulbs are very large and should be handled with gloves.

'Apple Blossom' has striking, funnel-shaped flowers on top of sturdy stems.

CHOOSE

When buying amaryllis bulbs, choose the largest ones you can find as these will produce the biggest and most plentiful blooms – sometimes two or three stems per bulb and up to three flowers per stem. Try to select bulbs that are fat, heavy, and dry.

One of the best species to look out for is the sturdy *Hippeastrum papilio*, which has gorgeous, green-hued white flowers with dark red markings. The most widely available amaryllis include the large-flowered types of the Galaxy Group, with their classic, trumpet-shaped blooms. 'Christmas Gift' has cream petals with a green base. Pink-washed 'Apple Blossom' and scarlet 'Red Lion' are also popular. 'Elvas' is a double-flowered cultivar.

Those from the *H. cybister* or Spider Group have more slender flowers with long, narrow petals. 'Merengue' is one of the finest, with up to three stems per bulb and up to six small red and green flowers per stem. 'Evergreen' has unusual, bright, green-flushed blooms.

PLANT AND CARE

Amaryllis is frost-tender and therefore best grown indoors in cool temperate climates. Grow it in a pot in a warm, bright position out of draughts and direct sunlight. Plant in early autumn for winter flowers, or any time up to midwinter for spring flowers. Blooms appear six to eight weeks after planting.

Before planting, soak the bulb's roots in tepid water for a few hours. Choose a pot that's only slightly bigger than the bulb, as amaryllis like to be confined and flower better when pot-bound. Mix up a rich, well-drained growing medium of one-third horticultural grit and two-thirds bulb or multipurpose compost. Half fill the pot with this mix. Set in the bulb, making sure the top is sitting above the rim of the pot; fill in with growing medium and firm in well. The bulb should have its top third out of the pot and the bottom two-thirds in the compost. Top with a layer of grit, then water and place in a warm, bright spot for a few weeks. As shoots grow, rotate the pot every few days and feed regularly with a balanced liquid fertilizer. Water regularly when in flower to stop the compost drying out. Varieties with big blooms may need staking.

With the proper care, you can encourage your amaryllis bulbs to reflower the following year (see p.43).

PROPAGATE Mature amaryllis (older than two years) produce offset bulblets. Remove these carefully from the main bulbs and grow them on in individual pots. They should flower in two years.

WAYS TO GROW

Amaryllis look great as part of a forced bulbs display on a window sill. Time it right, however, and you can create an amazing table centrepiece for Christmas by planting several bulbs together in a bowl planter.

DWARF IRIS *IRIS RETICULATA*

Dwarf irises are early-flowering bulbs with intricate, short-stemmed flowers that look like butterflies. They come in shades of pale to violet blue, purple, and dark red, or occasionally yellow, brown, and speckled. The orchid-like blooms are often fragrant and last up to three weeks.

BULB TYPE Bulb
HARDY Fully hardy
HEIGHT Up to 15cm (6in)
LEAF Upright, grass-like, grows longer after flowering; deciduous
POSITION Full sun, part shade
WARNING! Ingestion may cause severe discomfort. Wear gloves and wash hands after contact.

CHOOSE

Iris reticulata are miniature marvels with complex blooms. They comprise three inner standard petals, which are upright, and three hanging outer petals. The outer petals are known as "falls", and usually feature striking markings in yellow and white.

The most well-known cultivar is 'Harmony', which has fragrant royal-blue flowers with white-striped falls and a blaze of yellow on the crest.

'Alida' – a soft sky-blue with yellow splashes on the falls – is a newer introduction, lauded for its strong growth habit. Pale 'Katharine Hodgkin' stands out with its mix of white, yellow, and blue-green markings. It's more reliably perennial than other dwarf irises, is nicely scented, and flowers early. 'Pauline' flowers a little later, bearing deep, velvet-like purple flowers with flashes of white on its falls. 'J.S. Dijt' is reddish-purple with an orange crest and white splotches on the falls.

PLANT AND CARE

Dwarf irises must have well-drained soil and will rot in heavy or wet soils. They need moisture in spring but dry conditions during their dormant period in summer and autumn. The bulbs will grow in part shade, but choose a sunny position for the best results.

Plant dwarf iris bulbs in early autumn. In the garden, plant them 10cm (4in) deep, pointed end facing up, and spaced 8cm (3in) apart. If growing them in pots, top with a layer of horticultural grit to prevent the wet soil from splashing up and marking the petals when it rains.

The foliage continues to grow after flowering, reaching lengths of up to 30cm (12in). Allow the leaves to yellow and die back before removing them.

Dwarf irises are relatively trouble-free, but birds may damage petals. Although perennial, they often don't bloom in their second year, but may do so in subsequent years. This is because,

Grow 'J.S. Dijt' in a bulb bowl to enjoy its glorious scent and flowers.

WAYS TO GROW

Dwarf irises look great in groups of 10–15 at the front of a border or in rock and gravel gardens, where they're visible. Raise them up so the scent can be enjoyed in a bulb tray or bowl. They can be brought or forced indoors but the blooms don't last as long in warmer conditions.

each time they flower, the bulbs can split, making them too small to bloom the next year. To guarantee flowering, plant new bulbs each autumn.

PROPAGATE Clumps can be lifted and divided in early autumn. This will add to your stocks and ease the congestion that can prevent irises from flowering. The split, replanted bulbs may take a year or more to flower again.

'Katharine Hodgkin' has intricate flecks and veining in blue and yellow.

DAFFODIL *NARCISSUS*

The kings of spring, daffodils bring a sunny outlook to any garden with their jolly, golden-yellow blooms. These plants are instantly recognizable by their flowers, which have a central cup ringed with petals. Easy and dependable, they're reliably perennial and long-lived, increasing over the years.

BULB TYPE Bulb
HARDY Most fully hardy, some tender
HEIGHT 15–75cm (6–30in)
LEAF Narrow, long; deciduous
POSITION Full sun
WARNING! All parts are toxic. Ingestion may cause severe stomach discomfort; contact may irritate the skin. Wear gloves and wash hands after contact.

'Mount Hood' (D1) blooms open in a creamy yellow colour then fade to white.

CHOOSE

With up to 200 species and more than 30,000 cultivars, there's a vast range of daffodils to choose from. They come in many shapes and sizes, from miniature to tall trumpet-flowered types. The best known are white and yellow but a few have orange or pink markings. While the majority are hardy and grown outdoors, some Tazetta narcissus, such as the famous scented paperwhites, are grown indoors (see pp.42–43).

The choice is so large that daffodil cultivars have been grouped into 13 horticultural divisions (see right), based on their flower forms. This system makes it far easier to identify and select daffodils with the characteristics you're looking for when buying them.

DIVISION 1 The Trumpet daffodils include 'Mount Hood' and 'W.P. Milner'.
DIVISION 2 The Large Cup group includes 'Ice Follies' and 'Carlton'. Both these cultivars have the classic daffodil shape of a long ruffle-ended cup and a collar of flat triangular petals around it. They have one bloom per stem.
DIVISION 3 Small Cup daffodils feature a short central corona.
DIVISION 4 The Double type daffodils, such as 'Bridal Crown', have a ruffle of petals at the centre instead of a cup.
DIVISION 5 Triandrus hybrids, such as 'Hawera', have reflexed petals and more than one bloom per stem.
DIVISION 6 The Cyclamineus cultivars, including 'Jetfire', are usually small, with a downward-pointing cup and reflexed outer petals.

DIVISION 7 The Jonquilla daffodils are highly scented, with star-like petals and several blooms per stem.
DIVISION 8 Tazetta cultivars, such as 'Geranium' and 'Minnow', are fragrant and have lots of flowers per stem.
DIVISION 9 Poeticus daffodils are usually the last to flower, with one bloom per stem. They have a short, flat, disc-like corona that's often orange or red-rimmed.
DIVISION 10 The Bulbocodium types flower early in the season and have large funnel cups like hoop petticoats.
DIVISION 11 This group is formed of the Split Corona daffodils.
DIVISION 12 These are Miscellaneous cultivars that don't fit in other groups.
DIVISION 13 The final group includes the species and wild daffodils.

'Bridal Crown' (D4) has sweetly scented double flower clusters.

Dwarf 'Hawera' (D5) has multiple stems, each topped with several blooms.

'Geranium' (D8), like other Tazetta daffodils, makes an excellent cut flower.

PLANT AND CARE

Daffodils will grow in a broad range of situations but prefer a spot in full sun with well-drained soil. Plant your bulbs in autumn, at a depth of 15–18cm (6–7in), with the pointed end up. Space them at least 7cm (2¾in) apart. In containers, for a strong show, space them 5cm (2in) apart. Water well.

Narcissus bulbocodium **(D13)** is small but has large, lemon-yellow cups.

Leave the foliage to die back naturally after flowering, no matter how untidy it becomes. Deadhead the flowers or cut the whole flower stem off as they finish blooming. This will help make the bulbs stronger for the following year by preventing them from putting energy into setting seed. The exception to this is species daffodils that you wish to multiply by self-seeding. Allow

these to form seedheads and disperse their seed in late spring and early summer before cutting them back.

From the second year onwards, feed daffodils in spring – from when the foliage appears to when it starts to turn yellow – with a general fertilizer. Apply a mulch during the autumn to prevent them drying out.

Daffodils are prone to different pests and diseases, including narcissus bulb fly and narcissus eelworm. They're also susceptible to rots such as narcissus basal rot if grown in damp conditions (see p.29). Affected plants should be dug up and destroyed. A condition known as daffodil blindness may cause bulbs to decline and produce leaves but no flowers. Dividing congested clumps may help encourage flowering to start.

PROPAGATE Daffodils multiply well on their own over time, but they can be lifted and divided in late summer. Species types will self-seed if permitted. You can also propagate daffodils by chipping the bulbs (see p.26).

> **TOP TIP** DAFFODILS PREFER A SUNNY SPOT BUT A FEW WILL GROW AND FLOWER WELL IN LIGHT SHADE, SUCH AS 'JACK SNIPE' (D6) AND THE ELEGANT, LATE-FLOWERING 'ACTAEA' (D9).

WAYS TO GROW

Daffodils are extremely versatile plants and can be grown in beds, pots, and grass. Wherever you grow them, it's best to plant them in clumps or large groups rather than individually. The shorter cultivars such as 'Tête-à-tête' (D6) work well at the front of borders and in pots, where they'll be most visible. Many daffodils, such as 'February Gold' (D6), are excellent for naturalizing in a lawn.

Grow different daffodils in long grass for a show that lasts all spring.

SEASONAL SCHEMES WINTER AND EARLY SPRING

Bulbs that bloom in winter and early spring pair up brilliantly, and most will grow happily together as they like the same conditions and situation. Early-flowering bulbs also make excellent combinations with trees, shrubs, and perennials that add interesting features to the garden throughout the coldest months, from fabulous scented flowers to vibrant stems. Use them together in different ways to create delightful seasonal displays packed with colour and scent.

MAGICAL CARPETS

Some of the best companions for early-flowering bulbs are, quite simply, each other. It would be hard to find a more uplifting sight at this time of year than the carpets of colour created by low-growing bulbs blooming happily in tandem. They make for a striking groundcover beneath bare trees on a woodland walk. Plant the bulbs in large clumps and they'll spread and naturalize, making drifts of flowers to banish grey winter days. Christmas rose, *Helleborus niger*, makes a terrific perennial partner to this scheme, but avoid growing large evergreens like conifers in the same area as they'll cast too much shade and suck up all the moisture (see p.38).

RECREATE IT Swathes of delicate white snowdrops (*Galanthus nivalis*) **(1)** mingle with pockets of yellow winter aconite (*Eranthis hyemalis*) **(2)** and pink Eastern cyclamen (*Cyclamen coum*) **(3)**.

TOP TIP LOW-GROWING EARLY SPRING BULBS LOOK WONDERFUL BENEATH SMALL SHRUBS THAT FLOWER AT THIS TIME OF YEAR, MANY OF WHICH HAVE SCENTED BLOOMS. ENJOY THE BEST OF THE SEASON BY GROWING BULBS SUCH AS GRAPE HYACINTHS, SNOWDROPS, OR DWARF IRISES WITH FRAGRANT WINTER-FLOWERING SHRUBS SUCH AS DAPHNE, VIBURNUM, WINTERSWEET, OR SWEET BOX.

STUNNING STEMS

Many early bulbs are suited to growing beneath deciduous trees and shrubs, where they receive late winter and early spring sunshine, before the canopy above fills in. Exploit this by mixing them with trees and shrubs that have interesting bark and winter stems, for a burst of colour. Birch trees, paperbark maple, and Tibetan cherry are popular for pale or peeling bark, which contrasts with dark soil and brightly coloured bulbs. Shrubs such as dogwoods and willows add warm and fiery shades with their winter stems, which make the perfect backdrop.

RECREATE IT Eye-catching stems of dogwood (*Cornus sanguinea* 'Midwinter Fire') **(1)** combine here with yellow daffodils (*Narcissus* 'February Gold') **(2)**, the bark and stems of a silver birch tree **(3)**, bright blue spikes of Siberian squill (*Scilla sibirica*) **(4)**, and deep-red heather (*Erica* × *darleyensis* 'Kramer's Rote') **(5)**.

POTS OF PLENTY

You can easily create ingenious plant displays by hanging small pots on a slatted fence with hooks and twine or ribbon (*see left*). Perennials that flower during the early spring – primroses and violas, for example – make a pretty picture with blue- and purple-flowered bulbs, such as dwarf irises and crocuses. Then use daisy-like *Anemone blanda* to complete the scheme, which combines beautifully with the foliage of variegated ivy. Consider framing an arrangement like this on a fence opposite a window, to brighten up the view from inside.

RECREATE IT This display of striking blue blooms combines *Iris reticulata* 'Harmony' **(1)**, *Crocus* 'Blue Bird' **(2)**, *Anemone blanda* 'Blue Star' **(3)**, *Primula* 'Blue Denim' **(4)**, and violas **(5)**.

Using only bulbs, this stunning mid-to late-spring scheme combines blue grape hyacinths and windflowers with tulips such as 'Lilac Wonder' and 'Queen of Night' in shades of pink and purple. Allium buds and fritillary foliage hint at further stages to the display.

MID- TO LATE SPRING

As the days grow longer and warmer, the garden bursts to life with fresh shoots and buds, unfurling leaves, and an array of exciting bulbs flowering in profusion. This is the season to enjoy much-loved garden favourites such as tulips and hyacinths as well as wild bluebells and fritillaries, and delicate woodland beauties, like dog's tooth violets and wake robins. Plan for a succession of blooms – blue grape hyacinths and white snowflakes, for example, followed by blue camassias and white lily of the valley – and you can create abundant displays that will last for weeks with ease.

CAMASSIA *CAMASSIA*

The tall, upright flower spikes of camassias are densely packed with typically blue, but also purple, pink, or white, star-shaped blooms. They open like fireworks, from base to tip. With a penchant for moist soils and shade, they'll grow in places where many spring bulbs cannot.

BULB TYPE Bulb
HARDY Fully hardy
HEIGHT 50–150cm (20–60in)
LEAF Linear, strap-shaped; deciduous
POSITION Full sun, part shade

Camassia cusickii is a tall, elegant plant that grows well in grass.

CHOOSE

Long-flowering and long-lived, camassias come back reliably year after year. They're tough and hardy, with fresh green foliage appearing in late winter, followed by show-stopping flowers throughout spring. For a succession of blooms all season, plant types that flower one after the other.

One of the first to flower, *Camassia leichtlinii* subsp. *suksdorfii* Caerulea Group is also one of the tallest varieties, with sturdy stems topped with clear blue flowers. Next is violet-blue *C. cusickii*, followed by the most

widespread species, *C. quamash*, which is shorter and has wider flower spikes. It's less tolerant of waterlogged soils than other species. The Caerulea Group cultivar 'Maybelle' is the most compact and flowers later, with smaller, deep blue-purple flowerheads. It likes shelter but can manage heavier soils.

White-flowered types include *C. leichtlinii* subsp. *leichtlinii*, and the cultivar 'Sacajawea', with creamy-white blooms and cream-variegated leaves. 'Semiplena' has double white flowers.

PLANT AND CARE

Camassias love humus-rich, moisture-retentive soil in full sun but grow well in drier conditions and part shade. Some types tolerate wet winter soils as long as they don't sit in waterlogged, soggy conditions for a long period.

Plant bulbs in early autumn at a depth of about twice the size of the bulb with the pointed end facing up. Depending on the bulb, this will range from 10cm (4in) to 20cm (8in) deep. Space them 10–15cm (4–6in) apart. Water well and leave them to grow.

After flowering, remove the spent flower stems, but allow the foliage to yellow and die back completely so the bulb can store energy for next season. In exposed areas and colder climates, protect bulbs with a mulch in autumn.

If a mature clump fails to flower, it may need splitting. Camassias rarely suffer from diseases and pests, but might get munched by snails and slugs.

Plant camassia with the top, pointed end of the bulb facing upwards.

PROPAGATE Lift and divide large colonies of camassias in late summer when the bulbs are dormant. Either separate and replant them, spacing them further apart from one another, or take off the bulblets and pot them up individually to grow on. Seed can also be harvested when ripe in the late spring and early summer and sown into pots – however, it's unlikely they'll match their parent plant.

WAYS TO GROW

Camassias look fantastic naturalized in grass, but their leaves take a long time to die back, so they're more suitable for planting in long grass, such as wildflower meadows, than lawns. They also do well in bog and pond-side schemes, in containers, and tricky, damp, shady areas. These plants make excellent cut flowers.

LILY OF THE VALLEY

CONVALLARIA MAJALIS

This perennial beauty is hung with pure-white, bell-shaped blooms. The tubby little nodding flowers dangle daintily from one side of the arching stems, contrasting with the long, broad, dark-green leaves. Lily of the valley is famous for its exquisite scent and as a wonderful cut flower.

BULB TYPE Rhizome
HARDY Fully hardy
HEIGHT 10–50cm (4–20in)
LEAF Narrow, ovate; deciduous
POSITION Part shade, full shade, full sun
WARNING! Highly toxic – do not ingest. Wear gloves and wash hands after contact.

CHOOSE

Convallaria majalis, lily of the valley, is a deciduous woodland plant that returns each year and spreads rapidly in shady areas. If left to its own devices, it can overrun other plants and become invasive, although modern cultivars are considerably better behaved.

C. majalis var. *rosea* has pink-tinged flowers. 'Flore Pleno' is taller, to about 30cm (12in), and bears double flowers; white and pink forms are available. 'Prolificans' is an unusual double variety with many more flowers. The white 'Géant de Fortin' is the tallest, with large leaves and a strong scent, while 'Dorien' has some of the largest flower cups. Variegated types include 'Albostriata' and 'Albomarginata', with edging and stripes to the leaves.

PLANT AND CARE

Lily of the valley thrive in part shade but will tolerate full shade, and also full sun, as long as the soil is moist. Their ideal position is light or dappled shade in humus-rich, moisture-retentive soil.

These plants are usually available to buy as "bare roots" or rooted crowns in spring. Soak this rhizome in water for about 30 minutes. For best results, plant into individual pots of multi-purpose compost with the roots just under the surface. Grow on and plant out the following spring. When planting, place the roots or "pips" just beneath the soil surface and water well.

Water pot-grown plants throughout summer; apart from this, they're very low-maintenance. Allow the leaves to die back after flowering. Mulch with leaf mould or well-rotted garden compost in autumn. Lily of the valley can suffer from grey mould (see p.28).

PROPAGATE These plants multiply with ease. You can lift and divide congested colonies in autumn, pulling apart the clumps. Ensure each smaller section of roots you replant has visible pips – little bud-like bumps on the roots – and that they sit just below the soil surface. New pips will take two years to flower.

WAYS TO GROW

Lily of the valley makes an excellent groundcover for shady areas and north-facing gardens. It works well at the base of shrubs too (deciduous and evergreen). However, it will leave a gap when it dies back in summer, and may also spread, so it's not ideal for mixed borders.

The blooms of *C. majalis* **var. *rosea*** have hints of pink that deepen in shade.

'Albostriata' has yellow-striped foliage and spreads more slowly than other types.

BARRENWORT *EPIMEDIUM*

Also known as fairy wings or bishop's caps, barrenwort is a low-growing, shade-loving perennial. Its wiry stems are hung with airy, spurred flowers in orange, yellow, red, pink, or white, but it's grown just as much for its pretty foliage, which is often tinted or veined with red, copper, or pink.

BULB TYPE Rhizome
HARDY Fully hardy
HEIGHT 15–30cm (6–12in)
LEAF Heart-shaped; deciduous, semi-evergreen, evergreen. Spread 30cm–1m (12in–3ft 3in).
POSITION Part shade, full shade, full sun

CHOOSE

Barrenwort can be deciduous, semi-evergreen, or evergreen. *Epimedium × warleyense* is a compact evergreen type with heart-shaped leaves and attractive copper-orange and yellow flowers borne in clusters on wiry stems. 'Orangekönigin' is a highly recommended cultivar that doesn't spread as quickly. It has red-tinged leaves in winter.

The semi-evergreen or deciduous *E. grandiflorum* 'White Queen' keeps or loses its leaves depending on the temperatures during winter.

Epimedium × warleyense bears sprays of orange flowers with yellow centres.

E. × rubrum has bronze spring foliage that turns green, then red in autumn.

It has white flowers and maroon-flushed, arrow-shaped green foliage. *E. × youngianum* 'Niveum' is another good white cultivar. *E. × versicolor* 'Sulphureum', also semi-evergreen or deciduous, has pale yellow blooms that dangle from red flower stalks and dance above the burgundy, green-veined leaves. The robust *E. × perralchicum* 'Fröhnleiten' is lower-growing, to 20cm (8in), with yellow flowers and spiny green leaves. It's also semi-evergreen or deciduous.

The mostly deciduous *E. × rubrum,* known as red barrenwort, has dark pink flowers with creamy spurs.

PLANT AND CARE

"Bare root" rhizomes are available in spring. Soak in water for a few hours before placing in a pot of compost with the shoots just under the surface. Leave them in a sheltered spot outside to grow.

Potted plants can be planted out in autumn or early spring, spaced at least 30–38cm (12–15in) apart. Some barrenwort, such as *E. × rubrum* and *E. × versicolor*, withstand dry conditions; others, like *E. grandiflorum*, prefer damp. All prefer a sheltered spot out of winds or hot sun. Barrenwort will grow in full shade but part or dappled shade is best, and it will tolerate full sun as long as it stays moist. Well-drained, humus-rich, fertile soil is ideal for this plant.

PROPAGATE Divide barrenwort in spring or autumn. Dig up the plants and gently pull the rhizomes apart, or cut them up with a sterile sharp knife (see p.27) and replant. In autumn, cut back a third of the leaves as well.

WAYS TO GROW

Barrenwort makes an excellent groundcover, spreading to form a dense mat that suppresses weeds. It's invaluable in low light and even deep shade, and is one of the few plants that can cope with difficult dry shade situations, such as at the base of a Leyland cypress hedge.

DOG'S TOOTH VIOLET

ERYTHRONIUM

BULB TYPE Bulb
HARDY Fully hardy
HEIGHT 20–40cm (8–16in)
LEAF Ovate, often mottled or marbled; deciduous
POSITION Part shade, full shade

Dog's tooth violets are elegant plants for shade. The lily-like flowers, in white, yellow, or pink, have reflexed petals that curve backwards and are held on slender stems above marbled foliage. They may appear diminutive and dainty, but these perennials are tough, vigorous, and low-maintenance.

CHOOSE

Dog's tooth violets get their name from the colour and shape of the bulb, which is white, pointed, and long, like a dog's tooth. These plants are often also known as trout lily or fawn lily for their mottled, tongue-shaped foliage.

The most famous species of dog's tooth violet is *Erythronium dens-canis*, which bears upright stems with one or more mauve-pink blooms, and brown and yellow rings at the base of the petals. The green leaves are marbled with bronze. It reaches 15–25cm (6–10in).

The hybrid type E. 'Pagoda' is widely available and easy to grow. It has lemon-yellow flowers and reddish-bronze mottling on the foliage. 'Sundisc' is another yellow form and produces three to four flowers per stem.

'Kinfauns Sunrise' has pink flowers with yellow centres and white veining on the leaves. 'Apple Blossom' is a taller cultivar, to 30cm (12in), with white flowers marked with yellow and red. Two of the best white types are E. *californicum* 'White Beauty' and 'Brocklamont Inheritance'.

PLANT AND CARE

The ideal situation for dog's tooth violets is one that mimics their native woodlands. They flower best in part shade but will grow well in full shade too, and prefer a soil that stays cool and moist all year. The bulbs are prone to drying out so are often sent out packed in a damp wrap. Alternatively, you can buy potted plants. Plant the bulbs in autumn, in moist but well-drained, fertile, humus-rich soil. Set them 15cm (6in) deep, spaced 15cm (6in) apart.

As long as dog's tooth violets are planted in the right conditions, they don't need much attention and prefer to be undisturbed. Simply water them

Erythronium 'Kinfauns Sunrise' is a pretty pink hybrid that multiplies well.

WAYS TO GROW

As well as woodland schemes, dog's tooth violets work brilliantly planted in drifts in damp or spring bulb meadows. They'll light up shady borders and can grow in difficult areas such as in pots in a shaded side return or courtyard.

E. californicum 'Brocklamont Inheritance' can reach 40cm (15in).

if dry and allow the foliage to yellow and die back on its own so the bulb can store energy for next season.

PROPAGATE Lift and divide dog's tooth violets in late spring, after flowering, when you can also take off any offset bulblets to grow on in pots of compost. They don't like to be moved, so leave them for at least three or four years after planting and between the dividing and transplanting sessions.

FRITILLARY *FRITILLARIA*

There is a broad range of fritillaries, from the diminutive patterned snake's head fritillary, thought of as a native wildflower, to the statuesque crown imperial. They also vary in growing conditions, from damp to dry, but most are hardy, with bell-shaped flowers, and go dormant in summer.

BULB TYPE Bulb
HARDY Fully hardy
HEIGHT Up to 1.2m (4ft)
LEAF Narrow and glaucous or whorled; deciduous
POSITION Full sun, part shade
WARNING! Ingestion may cause stomach upset; contact may irritate the skin. Wear gloves when handling.

Fritillaria imperialis **'Rubra Maxima'** are unusually tall for spring bulbs.

F. raddeana looks delicate but is a robust bulb that grows to 50cm (20in).

The cultivar *F. uva-vulpis* 'Kew Form' blooms earlier than the species.

CHOOSE

The earliest species of fritillary to bloom is delicate *Fritillaria meleagris*, the snake's head fritillary. Small, with dainty nodding flowers, it's known for the tessellated, checkerboard or snakeskin-like markings on its petals – purple checked with white, or white checked with green. It grows to 38cm (15in) and is usually perennial, coming back each spring if it has the right conditions.

F. imperialis, the crown imperial, has big bulbs that produce a ring of bold orange flowers atop tall, sturdy stems,

crowned by a spiky tuft of leaf-like green bracts. It flowers later and is often grown as an annual, but will come back each year in the right conditions. All parts of the plant are said to have a rather unpleasant smell. 'Lutea' is a yellow-flowered form.

The unusual and exciting *F. raddeana* is a smaller version of the crown imperial, with stunning pale yellowish-green flowers. Other less commonly grown but widely available fritillaries for the garden include the low-growing *F. michailovskyi*, Michailovski's fritillary, which has brown and green flowers

dipped in yellow; *F. elwesii*, Lebanese fritillary, which is green with a black stripe and two or three flowerheads per stem; and *F. uva-vulpis*, fox's grape fritillary, which is dark purple, tipped with bright yellow.

F. persica, Persian lily, bears upright flower spikes that are covered in up to 30 deep purple flowers all along the stem, and reaches up to 1m (3ft). Good cultivars include 'Twin Towers Tribute', which produces two flower spikes per bulb; and 'Ivory Bells', which has creamy-white to green blooms and silvery leaves.

PLANT AND CARE

In the wild, fritillaries grow in diverse areas, from mountains to meadows, in vastly different conditions. However, all have bulbs that are planted in autumn, generally at a depth of three to four times the size of the bulb.

Snake's head fritillaries like damp grassland or meadow conditions, or that of woodland, in part shade or light shade. They grow fine in full sun, as long as they stay cool and moist (not waterlogged), through the year. Plant these fragile bulbs gently, in fertile, humus-rich, moist, but well-drained soil 10cm (4in) deep and 15cm (6in) apart.

Crown imperials need almost the opposite situation – dry and very well-drained, in full sun. Plant at least 30cm (12in) deep and 30cm (12in) apart. The bulbs have a hole or dip where the previous year's flower stem grew – this is the top, and should be planted facing up. Bulbs are prone to rot from moisture ingress, so if possible add sand, grit, or crushed stone to the hole at planting time to aid drainage.

F. michailovskyi, *F. elwesii*, *F. uva-vulpis*, *F. persica*, and *F. raddeana* all like quite similar conditions, in full sun and well-drained soil, and can tolerate

TOP TIP A PROPORTION OF FRITILLARY BULBS ARE LIKELY TO BE "BLIND" AND NOT FLOWER, FOR UNKNOWN REASONS, SO ALWAYS PLANT MORE THAN YOU NEED.

light shade and dry soil. Plant the bulbs on their side, 15cm (6in) deep and 15cm (6in) apart, or three to four times the size of the bulb, if larger.

If your garden soil is heavy – clay, for example – grow your bulbs in pots of compost with added grit. While dormant, keep them dry and protected from summer heat in light shade.

Leave the foliage of all types of fritillary to wither and die back on its own. If you don't want the bulbs to self-seed, deadhead them or remove the flower stem after flowering to divert energy to the bulb instead of to seed production.

Protect fresh growth from slugs and snails in spring. Plants are also prone to attack from lily beetle, which can be picked off the plants as you see them.

PROPAGATE Snake's head fritillaries will self-seed around and naturalize well, so don't deadhead the flowers and let the plant die back on its own. Lift and split established clumps in

F. persica **'Ivory Bells'** is pale green with masses of blooms along the stem.

early summer and, at the same time, remove any offsets to pot up and grow on. You can also do this for the bulbils of established crown imperials in late summer, but note that they usually don't like to be disturbed.

WAYS TO GROW

Snake's head fritillaries are excellent for naturalizing in lawns and spring bulb meadows or for growing in containers, as long as they're kept moist. They look great in borders, but mustn't be swamped by other plants. Fritillaries that prefer drier conditions work well in gravel and dry garden schemes, rock gardens, and in mixed borders, if well drained. On heavy soils, raised beds can help to offer the drainage they need.

Snake's head fritillaries naturalize well in grass such as lawns and meadows.

ENGLISH BLUEBELL

HYACINTHOIDES NON-SCRIPTA

The English bluebell is a native perennial wild flower known for its arching stems lined with nodding, blue, bell-shaped blooms. It often grows en masse, making carpets of blue in woodlands, hedgerows, and fields. A wonderful spring-nectar source for pollinators, the flowers are also lightly scented.

BULB TYPE Bulb
HARDY Fully hardy
HEIGHT 40cm (16in)
LEAF Linear, strap-shaped; deciduous
POSITION Part shade
WARNING! Ingestion may cause severe stomach upset. Wear gloves and wash hands after contact.

CHOOSE

The English bluebell has narrow leaves and a drooping flowerhead. The blooms, often violet-blue but occasionally white or pink, hang to one side. The flowers have a fruity scent and creamy anthers and pollen.

However, take care not to mistake it for another plant that's often sold as "bluebell" – *H. hispanica*, the Spanish bluebell, which originally escaped from garden plantings. Its cross-breeding with the wild English bluebell has led to fears that the latter, an important native plant, may become displaced. Spanish bluebells have broader leaves and are taller, at up to 60cm (24in). The flowers are unscented, paler blue, and arranged all around the more upright stem. The anthers and pollen are usually blue.

It's essential to source the true *H. non-scripta* from a reputable supplier, both to support the survival of this beautiful bloom and ensure your plants haven't been taken from the wild.

PLANT AND CARE

As woodland plants, bluebells like a humus-rich, moist but well-drained soil. They prefer to be in part shade, for example under deciduous trees, where they'll receive some sunlight early in the year, before the canopy fills out. Plant dry bulbs in autumn, 10–15cm (4–6in)

English bluebells droop elegantly to one side of the plant.

Don't mistake Spanish bluebells for English bluebells when buying.

WAYS TO GROW

English bluebells are great for naturalizing in grass, but more suitable for meadow-style plantings than lawns, as the foliage doesn't die back until late in the year. They're also a top choice for shady garden borders and under deciduous trees and shrubs, and in large drifts in woodland gardens.

deep and 15cm (6in) apart, pointed end facing up. For best results, however, source bluebells from cultivated stock "in the green" (see p.18), freshly lifted while still in growth in spring. Place at the same depth as they were growing before – look for where the foliage goes white towards the base. Water well after planting. Allow a few seasons for plants to re-establish and flower. If the foliage is damaged, it can take years to recover. Don't step on the leaves or cut the foliage down.

PROPAGATE Bluebells multiply well by self-seeding. They can also be lifted and divided in summer, but may not flower the next spring while they re-establish. You can collect seed to sow from your garden plants in late summer, but it can take up to seven years for a flowering plant to develop from a seed.

HYACINTH
HYACINTHUS ORIENTALIS

Hyacinths are intensely fragrant spring bulbs with large, tubular flower-spikes of waxy, densely packed, bell-shaped blooms in shades of blue, purple, pink, white, red, or yellow. They're easy to grow and make for a bold, colourful display indoors or in the garden.

BULB TYPE Bulb
HARDY Fully hardy
HEIGHT Up to 30cm (12in)
LEAF Upright, broad, strap-shaped; deciduous
POSITION Full sun, part shade
WARNING! Ingestion may cause severe discomfort; contact with bulbs can irritate the skin. Wear gloves and wash hands after contact.

Dutch hyacinths are bold and bright, with large, super-scented flowers.

CHOOSE

Hyacinthus orientalis, often known as Dutch hyacinth, is the most commonly available type of hyacinth, with a huge number of super-scented cultivars in a rainbow of colours. 'Carnegie' is a good, pure white variety. 'Delft Blue' has sky-blue flowers, while 'Blue Jacket' blooms are a deeper purple-blue.

For pink-flowered hyacinths, look out for 'Pink Pearl', which has dark pink centres and pale pink tips. 'Woodstock' is an interesting dark plum to reddish-purple. More unusual colours include the pale yellow 'City of Haarlem' and apricot-pink 'Gipsy Queen'. The flamboyant double-flowered types include peach-toned 'Annabelle' and the white 'Madame Sophie'.

Hyacinthus orientalis var. *albulus*, Roman hyacinths, are more natural looking, with looser flower spikes, but they're less widely available. For a similar look, seek out types sold as Multiflora hyacinths, which have multiple stems per bulb of more open, relaxed-looking flowerheads – such as *H. orientalis* 'Anastasia' or the varieties 'White Festival' and 'Blue Festival'.

PLANT AND CARE

When sourcing hyacinths, you'll find the same cultivars are available in two different types of bulb: one is for outdoor flowers in spring, while the other has been specially prepared to grow inside and flower earlier, in winter.

If growing outside, plant your bulbs in autumn in well-drained, reasonably fertile soil, in a sunny spot. They'll tolerate partial shade for one season only. Place them 10–15cm (4–6in) deep and 10cm (4in) apart, with the pointier end facing up. Water well after planting.

In containers, add grit to the compost for good drainage. Pot-grown hyacinths tend not to be as hardy as those in the ground, so they may need protection from frost. Feed with liquid tomato fertilizer in the growing season.

You can force prepared bulbs inside in early autumn (see p.42). Alternatively, bring your pot-grown, spring-flowering hyacinths inside as they start to bloom.

Taller cultivars with large flowers may need staking. Remove the spent flower spikes but let the foliage die back. If

WAYS TO GROW

Hyacinths are big, showy flowers that lend themselves to bold, bright displays with other spring bulbs such as tulips and daffodils, in borders or pots. They're often grown or brought indoors to conservatories or window sills for their strong, heady fragrance.

Hyacinths can be forced indoors in bulb vases with water.

your soil is well-drained, leave the bulbs in the ground. Otherwise, it's worth lifting and storing them in summer (see p.23) until replanting in autumn. They are perennials, but may flower less or disappear after a couple of years, so replace or add to displays every season.

PROPAGATE Lift your hyacinth plants in late summer and remove offsets to pot up in compost and grow on. You can also propagate by scaling (see p.25) or chipping the bulbs (see p.26).

SPRING STARFLOWER

IPHEION UNIFLORUM

BULB TYPE Bulb
HARDY Fully hardy
HEIGHT Up to 25cm (10in)
LEAF Narrow, strap-shaped; deciduous
POSITION Full sun, part shade

Single, perfect star-shaped blooms adorn this easy-care perennial, which isn't as widely grown as it deserves to be. Spring starflower has lax, pale green leaves that smell faintly oniony, followed by honey-scented flowers with pointed petals in colours from white and pale silvery-blue to lilac and pink.

Ipheion uniflorum is a low-care, easy-to-grow perennial bulb.

Choose *I.* **'Rolf Fiedler'** if you want the darkest blue cultivar.

CHOOSE

Spring starflower has variably silver-white to pale-blue tinged blooms, often with a darker line or stripe down the centre of the petals. The grass-like foliage comes first, sometimes in autumn but most often in winter. In spring, short upright stems appear, each one bearing a starry flower.

The cultivar 'Alberto Castillo' is one of the tallest and earliest-flowering hybrids. Its white flowers have a central green vein. 'Charlotte Bishop' is another early flowerer, with light pink blooms.

One of the most popular varieties is 'Rolf Fiedler', which has blue flowers with a deeper blue central stripe. Paler

'Wisley Blue' has lilac-blue flowers and is shorter, to just 10cm (4in). 'Froyle Mill', 15cm (6in), has violet flowers.

PLANT AND CARE

Spring starflower prefers a warm, sheltered spot in full sun, but will do best with protection from baking heat on summer afternoons. Its ideal spot would mimic the conditions of a woodland edge, with part shade and some moisture early in the season.

Plant in early autumn in fertile, moist but well-drained soil, improving with organic matter if necessary. Dig planting holes 8cm (3in) deep and place bulbs at least 5cm (2in) apart. Water well.

These plants are low-maintenance and require little care or attention once established. Don't remove foliage as it yellows – let it die back naturally to soak up light and store energy for next year. Leave the bulbs in the ground for a show year after year. Mulch in autumn to protect the bulbs from frost. In late winter and early spring, watch out for slug and snail damage.

Spring starflower will spread happily in the right conditions but can become invasive in milder climates. If desired, deadhead the spent flowers so that they can't self-seed, and dig up and divide clumps every year.

PROPAGATE Lift and divide the bulbs in summer when dormant, pulling apart established clumps and replanting the smaller groups of bulbs immediately.

WAYS TO GROW

Spring starflowers will be content planted at the feet of deciduous shrubs, including roses. They're short, so work best at the front of borders or with other small plants in alpine troughs and rock gardens. They act as a good groundcover option for suppressing weeds in an area from winter to summer, and make excellent cut flowers.

BLEEDING HEART *LAMPROCAPNOS*

As its name suggests, bleeding heart has heart-shaped flowers that dangle from long, arching stems. Below each heart hangs an intriguing teardrop-like petal. Delicate foliage appears in spring, creating a large clump that dies back over summer, but returns stronger each spring.

BULB TYPE Rhizome
HARDY Fully hardy
HEIGHT 50–90cm (20–36in)
LEAF Divided; deciduous. Spreads 50cm (20in) or more.
POSITION Part shade, full sun
WARNING! All parts are toxic. Ingestion may cause severe stomach discomfort; contact may irritate the skin. Wear gloves and wash hands after contact.

CHOOSE

Bleeding heart is also known as lady in the bath and Dutchman's breeches. Its botanical name was changed from *Dicentra* to *Lamprocapnos*, though a few plants remain in the former genus. The species *L. spectabilis* is widely available, and bears up to 10 or 11 dark rose-pink blooms along its graceful stems. In each intricate flower, there are two pink outer petals with little spurs at the bottom, which curve out at the edges; the two inner, white petals form a teardrop. The cultivar 'Alba' has all-white flowers and tends to grow slightly taller, up to 1m (3ft). 'Gold Heart' is an unusual variety with green foliage in sun but golden-yellow leaves when grown in the shade.

L. spectabilis VALENTINE is a recent introduction with cherry-orange blooms. It's more compact and has pretty young foliage that's tinged purple.

Lamprocapnos spectabilis 'Alba' has pendent flowers on arching stems.

Dicentra formosa 'Bacchanal' has dark red blooms over a long period.

Keeping the old name, *Dicentra formosa* 'Bacchanal' is shorter, to around 50cm (20in), but with a wider, spreading habit to 1m (3ft). It has deep maroon flowers, which bloom through spring into summer, and grey-green leaves.

PLANT AND CARE

Bleeding hearts prefer part shade, but they'll tolerate a bright, sunny spot as long as they have moist soil. *Dicentra formosa* 'Bacchanal' is a little less hardy than *L. spectabilis* types and benefits from a more sheltered position out of hot sun and cold, drying winds in winter.

These plants need a moist, humus-rich soil that's neutral to slightly alkaline. It should be friable and not too heavy: if necessary, improve the soil by adding organic material such as leaf mould to the hole before planting.

Source bare-root rhizomes to plant in early spring, or pot plants to plant in spring or early summer. Place the bare roots 8cm (3in) deep, spaced at least 40cm (16in) apart. Water well.

Deadhead spent flowers, first to encourage more, and then to stop the plant from putting energy into seed production. Allow foliage to die back. Mulch in autumn. Feed with general purpose fertilizer in the growing season.

PROPAGATE Bleeding hearts don't like disturbance, so instead of dividing, take root cuttings in late autumn. Remove the soil from around one section of roots. Cut off a piece of thick root about 4cm (1½in) long, and cover the rest back with soil. Place the cutting on its side in a fresh pot of compost, just covered, and keep moist until roots form. Leave the pot in a frost-free place over winter and plant out in the garden once leaves show in spring.

WAYS TO GROW

These perennials are a terrific choice for underplanting trees and shrubs, or for the middle of a shady border, with other plants around them that will fill in and cover the gap they leave later in the season. Bleeding hearts are not the best option for containers as they disappear completely in the summer, leaving a bare pot.

SNOWFLAKE *LEUCOJUM*

Snowflakes are often mistaken for their relatives, snowdrops (*see pp.52–53*), but are taller and flower later. They bloom for up to three weeks, producing arching stems topped with clusters of bell-shaped white flowers with pointed petals, which are tipped with green or occasionally yellow markings.

BULB TYPE Bulb
HARDY Fully hardy
HEIGHT 15–50cm (6–20in)
LEAF Upright, narrow, strap-shaped; deciduous
POSITION Part shade, full sun

Long-lived perennials, snowflakes come back reliably each spring.

CHOOSE

Two species of bulbous snowflakes are widely available to gardeners. *Leucojum vernum*, the spring snowflake, is the earlier to flower, and grows to around 30cm (12in), with one or two flowers per stem. *L. aestivum*, summer snowflake, blooms in mid-spring, despite its name. 'Gravetye Giant' –

which is the species' best-known cultivar – produces the tallest stems and plentiful flowers.

PLANT AND CARE

Snowflakes prefer part shade, but will grow in full sun in areas that don't get too hot and dry in summer. Plant bulbs in any moderately fertile, well-drained soil that's moist but not boggy, adding organic matter if the soil needs improving.

You can plant dry bulbs in autumn, but you'll get better results by buying bulbs "in the green" (*see p.18*). These are lifted in spring, just after flowering,

Spring snowflakes attract early pollinators, including bees.

WAYS TO GROW

Snowflakes are great for shade gardens, around trees, or near a pond, and will happily naturalize in a lawn, as long as the grass isn't cut until the leaves have died back. They look great in containers or planted near the front of a border, though taller-flowering varieties may need support.

and are ready for immediate planting. Space them 8–10cm (3–4in) apart and plant at a depth of 10cm (4in).

After flowering, allow the leaves to die down before cutting them back. Snowflakes will benefit from a mulch around the plants in autumn and potassium-rich fertilizer in the growing season (*see p.21*). Keep the soil moist, watering if necessary – snowflakes will fail if the soil dries out in summer. They may need protection from slugs and snails, which love eating the flowers, and can also suffer damage from narcissus bulb fly (*see p.28*).

PROPAGATE Over time, snowflakes will bulk up and slowly spread to fill the spaces between plants. They hate being disturbed, so you'll only need to lift clumps every five years or so to divide them – do this in autumn, when offsets can also be removed and replanted. Chipping and scoring bulbs are also options (*see p.26*).

GRAPE HYACINTH *MUSCARI*

Small, sweet, and very easy to grow, grape hyacinths bring a welcome pop of bright blue to the garden in spring. Their dense cone-shaped flower spikes, comprising crowded little clusters of florets, are reminiscent of an upside-down bunch of grapes. As an added bonus, many are also scented.

BULB TYPE Bulb
HARDY Fully hardy
HEIGHT Up to 30cm (12in)
LEAF Narrow, linear, grass-like; deciduous
POSITION Full sun, part shade

'Valerie Finnis' naturalizes well, spreading quickly to form large colonies.

CHOOSE

Muscari armeniacum, with its violet-blue, white-tipped flowers, is one of the most popular species of grape hyacinth. Look out for 'Valerie Finnis', a bright azure, fragrant cultivar with tightly packed flowers, and the free-flowering 'Blue Spike'. *M. neglectum*, considered a British native wildflower, has dark blue-black blooms with white tips and a paler lavender tuft. It freely self-seeds, so is not suitable for beds in small gardens. *M. azureum*, with soft pale blue blooms, is less vigorous and flowers earlier in the spring. *M. latifolium* is dark purple-blue below with a bright mauve-blue top. Another two-tone bloom is *M. macrocarpum* 'Golden Fragrance', an unusual, highly scented yellow form with purple tops.

PLANT AND CARE

Grape hyacinths thrive in full sun or part shade in well-drained or moist but well-drained soil. They're less vigorous in soils with lower fertility, such as stony sites or under trees, but will romp away in rich soil in borders. Plant bulbs in autumn, pointed end up, 10cm (4in) deep and spaced 8cm (3in) apart.

These plants hate soggy roots, so only water them if the soil becomes really dry. After flowering, remove flowerheads to prevent self-seeding, but leave the foliage on until it has turned yellow or died back. Some species, such as *M. armeniacum*, may send out fresh leaves in autumn. Don't worry about this – the leaves may die off in winter, but more leaves will appear with the flowers in the spring.

PROPAGATE To multiply plants, dig up, divide, and replant clumps in late spring, after flowering, with foliage attached, or in summer, when dormant. Offset bulblets can be removed at the same time and planted in pots of compost to grow on. Many types will spread and self-seed naturally. Their vigour is great for creating large, impactful displays.

WAYS TO GROW

Grape hyacinths look terrific in clusters of five or more bulbs, or can be mass planted by digging out a small area and planting multiple bulbs at once. They're fantastic for filling gaps towards the front of a border in spring, covering bare soil under deciduous shrubs, naturalizing in grass, or in rock gardens. The plants are also perfect for pots, on their own or in mixed plantings, and even thrive in small containers such as window boxes.

Grape hyacinths mix with violas, ferns, and grasses in this window box.

SOLOMON'S SEAL *POLYGONATUM*

A slow-growing but long-lived, low-maintenance perennial, Solomon's seal has slender, arching stems that carry small clusters of cream to white, nodding, bell-shaped flowers. The spring blooms are followed by black berries in late summer. Its smooth, dark-green leaves turn golden yellow in autumn.

BULB TYPE Rhizome

HARDY Fully hardy

HEIGHT Up to 1.2m (4ft)

LEAF Ovate or lance-shaped; deciduous. Spreads to 30–60cm (12–24in).

POSITION Part shade, full shade

WARNING! All parts are toxic, especially to pets. Ingestion may cause severe discomfort. Wear gloves and wash hands after contact.

CHOOSE

Polygonatum × *hybridum*, garden Solomon's seal, is the most commonly grown type. It's also one of the tallest, reaching 1.2m (4ft). Recommended cultivars tend to have foliage interest, such as the cream-streaked leaves of 'Striatum' or 'Betburg', which has purple-brown stems, young leaves, and grows to 75cm (30in).

Polygonatum odoratum, angular Solomon's seal, has single leaves along the stem, rather than pairs. *P. odoratum* var. *pluriflorum* 'Variegatum' has rounder leaves rimmed with white. Other cultivars include the double-flowered form 'Flore Pleno'; the eye-catching 'Red Stem'; and 'Silver Wings', which has blue-hued foliage and is shorter at around 40cm (16in).

P. verticillatum, whorled Solomon's seal, as its name suggests, has whorled leaves. The cultivar 'Rubrum' has pretty pink flowers.

Polygonatum is sometimes confused with *Maianthemum racemosum*, false Solomon's seal, which has similar leaves.

PLANT AND CARE

Solomon's seal is tolerant of most soils, providing they're humus-rich and moist but well drained. Its ideal position is somewhere damp as well as cool, shady, and sheltered.

WAYS TO GROW

Solomon's seals are excellent plants for woodland schemes with other shade lovers such as ferns. They also bring welcome height to the middle of a shaded border and act as good companion plants in cottage-garden plantings. The flowers and foliage are prized for cutting and bringing indoors for the vase.

Order bare-root rhizomes in late winter to plant out in early spring. First, soak the rhizome in a bucket of water for about an hour, then plant around 10cm (4in) deep and at least 15cm (6in) apart. Lay it horizontally in the planting hole, making sure that any shoots or stems remain above the soil, and any buds just beneath it. Cover the roots and firm the soil. Water well.

After flowering, leave the foliage until the autumn, when you can cut it back if you wish. Mulch in spring to keep it cool and moist.

Planted in the right place and given time to establish undisturbed, Solomon's seal needs little attention. Look out for damage from sawfly larvae in early summer, picking off any you see.

PROPAGATE Divide in early spring every three to four years. Lift and separate or cut up the rhizome into smaller pieces (see p.27), making sure each section has its own bud, before replanting.

Polygonatum × hybridum is a graceful plant for shady spring gardens.

STRIPED SQUILL *PUSCHKINIA*

Short but sweet, striped squill produce upright spikes covered with up to 10 pale, star-shaped blooms. The white flowers, which usually feature a blue stripe along the centre of each petal, are scented and good for pollinators, so they deserve to be more widely grown.

BULB TYPE Bulb
HARDY Fully hardy
HEIGHT Up to 20cm (8in)
LEAF Strap-shaped, upright; deciduous
POSITION Full sun, part shade

Puschkinia scilloides **var.** *libanotica* is easy and trouble-free, and will grow happily in sun or shade.

CHOOSE

Puschkinia scilloides var. *libanotica*, also known as Russian snowdrop, is easy to grow, hardy, and reliably perennial. It bears long-lasting spring flowers and is trouble-free, as well as deer and rodent resistant. The plant reaches 10–20cm (4–8in), depending on where it's grown, and bears upright, strap-like, bright green leaves, which appear at the same time as the blooms. The delicate white flowers appear from a distance as pale blue because of the darker line in the middle of the petals.

The bulbs are often sold as *P. scilloides* or *P. libanotica*. A pure white cultivar called 'Alba', with a lovely scent, is also available. Both the species and cultivar are nectar-rich and loved by early bees.

PLANT AND CARE

Striped squill will grow in both sun and part shade, but you'll get the best results if your plant can receive a bit of both – around the base of deciduous trees, for example, where it can enjoy moist conditions and sunlight in spring, but a drier summer – an ideal situation for developing the bulbs.

Plant in autumn in any moist but well-drained soil. Place bulbs with the pointed end facing up, at a depth of 10–15cm (4–6in) and spaced at least 10cm (4in) apart, as striped squill don't like being crowded. Water well.

Don't mow, cut, or remove the foliage until after it has yellowed and died back by itself. Mulch in the autumn. If the plants stop flowering, lift and divide them in late summer or autumn. Striped squill don't usually suffer from pests and diseases.

PROPAGATE These bulbs are good spreaders via self-seeding and increase quickly on their own. If you wish, you can divide them in late summer or autumn, when you can also remove offsets to grow on in pots of compost.

WAYS TO GROW

Striped squill naturalize well so are great for creating carpets of spring blooms, filling the gap between early flowers, like snowdrops, and later ones, like bluebells. They add colour under and around trees, hedges, and shrubs, or in rock gardens, and they also work well in containers.

P. scilloides **var.** *libanotica* 'Alba' is pure white, with no central stripe.

SQUILL *SCILLA*

Squills are a large group of mostly spring-flowering bulbs with clusters of star-shaped blue, pink, or white flowers arranged in various ways around short stems. Although tiny in stature, they pack a punch when planted en masse, and come back year after year, multiplying well over time.

BULB TYPE Bulb
HARDY Fully hardy
HEIGHT 15–30cm (6–12in)
LEAF Sword-shaped, arching; deciduous or evergreen
POSITION Full sun, part shade
WARNING! Ingestion may cause severe discomfort. Wear gloves and wash hands after contact.

Scilla forbesii is also called *Chionodoxa* and glory of the snow.

S. mischtschenkoana looks delicate but is hardy and early to bloom.

CHOOSE

Scilla forbesii, often sold as *Chionodoxa* and commonly known as glory of the snow, blooms early in spring. It bears clusters of deep blue, starry flowers with white centres, sometimes 12 on one stem. *Scilla* 'Blue Giant' is most reliable and taller, with larger flowers. 'Pink Giant' is pink with white centres.

S. siberica, Siberian squill, flowers almost as early. It grows to around 20cm (8in), with flower spikes carrying up to five nodding, cup-shaped, bright blue flowers. 'Spring Beauty' is larger, with deep blue flowers, and 'Alba' has white blooms. *S. mischtschenkoana*, Misczenko squill, is another early bloom, with flowers in a pale silver-blue tone. 'Tubergeniana' is the best-known cultivar.

Coming later in spring, and very different in appearance, is *S. peruviana*, Portuguese squill, which has evergreen to semi-evergreen foliage and stems to 30cm (12in), topped with large, conical flowerheads made up of unusual indigo-blue blooms.

PLANT AND CARE

Most squills are easy to grow and tolerant of a range of conditions. They'll grow in full sun or part shade, in any moderately fertile, humus-rich, well-drained soil. Plant in early autumn, 8–10cm (3–4in) deep, depending on the size of the bulb, and 8–10cm (3–4in) apart. The bulbs can dry out easily and prefer some moisture in autumn, winter, and spring, but can be drier in summer.

Avoid deep shade and heavy, waterlogged soils. Leave the foliage to turn yellow after flowering. If growing in grass, don't mow until the leaves have died back completely.

PROPAGATE Many squills will self-seed. You can also lift and divide clumps in autumn, removing offsets at the same time to pot up and grow on, if you wish. *S. peruviana* will not produce offsets as quickly as the others and should be left undisturbed for several years.

S. peruviana's large flowerheads open gradually, from the outside in.

WAYS TO GROW

Early squills such as glory of the snow look terrific grown in large swathes under trees with crocuses and snowdrops. They're a good choice for naturalizing in lawns as the foliage dies back quickly. Squills such as *S. peruviana* look great in pots and at the front of borders.

WAKE ROBIN *TRILLIUM*

These are showy plants for shade, with flame-like flowers in colours from yellow and white to dark red. Everything appears in threes, with three-petalled blooms opening above three green sepals, which emerge from a whorl of three pointed leaves. The foliage is often marked or mottled.

BULB TYPE Rhizome
HARDY Fully hardy
HEIGHT Up to 60cm (24in)
LEAF Diamond-shaped or oval; deciduous. Spreads to 30cm (12in).
POSITION Part shade, full shade

CHOOSE

Also called toadshade, trinity flower, and wood lily, wake robin grows wild in the forests of North America, but there are several types suitable for growing in gardens. *Trillium grandiflorum*, American wake robin, produces large clumps of stems to 40cm (16in), with plain leaves topped by white flowers. *T. grandiflorum* f. *roseum* has pink blooms, and there are two double-flowered white forms, *T. grandiflorum*
f. *polymerum* 'Snowbunting' and 'Flore Pleno'. *T. erectum*, birthroot, has dangling, purple-red flowers and plain green leaves. *T. chloropetalum* var. *giganteum*, giant wake robin, has darker maroon blooms and mottled, brown foliage. Both this plant and *T. erectum* are variable and can also bloom in other colours such as yellow, white, or purple. *T. cuneatum*, known as little sweet Betsy, has dark red flowers and mottled grey leaves like camouflage. *T. luteum*, yellow wood trillium, bears yellow flowers and silver-splashed, green foliage.

PLANT AND CARE

Wake robins have a reputation for being fussy, but they're trouble-free plants when happy, and will thrive and even become vigorous if given their

WAYS TO GROW

Wake robins are great groundcover for woodland schemes, though not right at the base of trees where it's dry. They're perfect for shady gardens, at the base of walls, or in the shadow of buildings with other plants that like the same conditions, such as barrenworts (see p.64) and dog's tooth violets (see p.65).

preferred conditions. As woodland plants, they like a sheltered, shady spot in fertile, humus-rich, acid to neutral soil that's moist but well-drained.

If buying wake robins in pots, plant them out in spring, once all risk of frost has passed. If buying rhizomes, make sure to source moist, dormant rhizomes, not dry ones. Plant them in autumn, about 5cm (2in) deep and spaced at least the same distance apart, laying the rhizome on its side. Water well and keep moist while the plant establishes. Once planted, they're best left undisturbed so they can settle in. Don't remove leaves, but allow to die back on their own. Mulch in autumn with leaf mould.

PROPAGATE Leave wake robins undisturbed for several years so they can bulk up. Divide after flowering, being careful to avoid damaging the roots when lifting. Given the right conditions they'll self-seed.

Trillium grandiflorum f. roseum blooms have three pale pink petals.

Place the rhizome on its side in the planting hole.

TULIP *TULIPA*

Nothing brings bright pops of colour to the spring garden quite like tulips, from candy pink and yellow to hot red and orange, pretty purples, and whites. These stunning, cup- or goblet-shaped blooms sit atop individual upright stems, and come in a variety of sizes, forms, and flowering times.

BULB TYPE Bulb
HARDY Fully hardy
HEIGHT 10–75cm (4–30in)
LEAF Strap-shaped; deciduous
POSITION Full sun
WARNING! Ingestion may cause stomach upset. Wear gloves and wash hands after contact.

CHOOSE

There are thousands of tulips to choose from, so to help classify them more easily, they've been split into groups. Some, such as Single Early, Double Early, Single Late, and Double Late, helpfully describe both the flower shape and flowering time. Many are grouped according to their flower forms, including Lily-flowered tulips, Parrot tulips (with ruffle-edged petals), and Fringed tulips. Other commonly available types include Viridiflora, Triumph, and Darwin hybrid tulips.

Tulips come in different sizes, from small species types of around 10–20cm (4–8in) to large, showy tulips that top 60cm (24in). They also have different blooming times, in early, mid-, or late spring. Some tulips are sold to grow in borders and pots for just one season, while others will come back every year, and naturalize well in grass, such as *Tulipa sylvestris*, wild tulip.

One of the shortest and earliest to flower is the species type *T. humilis*, to just 10cm (4in), and its cultivar 'Little Beauty', which is a real showstopper with hot pink petals and a blue centre.

Other worthwhile dwarf species and cultivars include mauve, yellow-centred *T. saxatilis* (Bakeri Group) 'Lilac Wonder', white, yellow-centred *T. tarda*, and bright red *T. sprengeri*.

Early border tulips include the unusual bi-colour *Tulipa* 'Prinses Irene', which has pointed orange petals with a maroon flame up the centre, and the best white, 'Purissima'. Good mid-season tulips include the soft red 'Couleur Cardinal', and 'Menton' in shades of apricot-pink. A dramatic late-season choice is double-flowered, blush-pink 'Angélique', which has

'Bright Parrot' has dramatic, ruffled petals in dusky red with yellow markings.

'Little Beauty' is small, with pretty pink petals and blue centres.

Orange **'Prinses Irene'** combines brilliantly with blue forget-me-nots.

'**Purissima**' is an elegant plant, with large, creamy-white blooms.

bowl-shaped blooms so opulent they could be mistaken for a peony. In contrast, the darkest tulips include 'Queen of Night', 'Havran', and 'Paul Scherer', all popular single, late varieties with deep purple, almost black flowers.

PLANT AND CARE

There are several tulips, such as 'Apricot Beauty', that will tolerate light shade, but in general they grow best in full sun in a sheltered spot with protection from strong winds. Most tulips like a neutral to alkaline, fertile, well-drained soil, but some species types – including *T. sylvestris*, *T. sprengeri*, and *T. tarda* – need a moist soil. Plant in mid- to late autumn. The planting depth and spacing will depend on the size of the bulb, but a good guide is to dig a hole about three times as deep as the bulb's height. If in doubt, plant deeper rather than shallower. Space at least two bulb widths apart.

Wild species types and their related cultivars are reliably perennial, as are some border tulips like 'Spring Green', a late-season, ivory and green tulip, and

rich pinky-purple 'Negrita'. Deadhead the spent flowers but allow the foliage to yellow and wither before removing.

However, most tulips won't come back every year and as a result are treated like annual bedding – used for flowers for one season, then lifted after blooming and discarded, preferably on the compost heap. Those in pots can be planted out in the ground afterwards to see if they reflower the next year.

Squirrels sometimes dig up freshly planted bulbs, but the biggest issues are diseases such as tulip fire, tulip viruses, mould, and rot (see pp.28–29).

PROPAGATE Perennial tulips can be propagated by division in summer by lifting and removing offset bulblets. Pot them up in compost to grow on or replant them at the original depth the bulbs were growing.

Late-season 'Angélique' has double, bowl-shaped blooms in shades of pink.

TOP TIP WHEN CREATING A MIXED TULIP DISPLAY, MAKE SURE YOU FIND OUT WHEN EACH CULTIVAR FLOWERS SO YOU CAN PLAN FOR YOUR COMBINATION TO ALL BLOOM AT THE SAME TIME.

Create a bright spring scheme with tulips in an array of candy colours.

WAYS TO GROW

Both species and wild tulips can be successfully naturalized in grass, but the other types are usually grown in borders, beds, and containers. The staggered flowering times of different tulips means you can create several schemes by planting early, mid-, and late types, perhaps with different colour palettes. Alternatively, carry the same shades throughout the season in a prolonged show, with one tulip taking over from another.

SEASONAL SCHEMES

This is a wonderful season to spend time in the garden, as the sun warms the soil and the days gradually lengthen. Lawns, borders, and containers burst forth with fresh new shoots and buds and everything comes suddenly to life, lush and lovely, with welcome pops of bright and cheery colour from pretty blooms such as tulips, hyacinths, and fritillaries. Although night frosts may still threaten tender young plants, there's a multitude of hardy bulbs to plant, grow, and enjoy.

SPRING ZING

Bring some zest to the second half of spring by combining tall, flowering bulbs like camassias or crown imperial fritillaries with plants with chartreuse foliage tones and bright acid-green flowers. Lime-yellow spurge is one of the best perennial plants for interest at this time of year. If you don't have room for this bushy subshrub, you can achieve the same effect with the flowers of lady's mantle, *Alchemilla mollis*, or, in the shade, with perfoliate alexander, *Smyrnium perfoliatum*.

RECREATE IT This fresh spring scheme brings together spurge (*Euphorbia palustris*) **(1)** with pots of bulbs, including bronze-red tulips (*Tulipa* 'Abu Hassan') **(2)** and daffodils (*Narcissus jonquilla* 'Flore Pleno') in galvanized buckets **(3)**. In the foreground, tall blue camassias (*Camassia leichtlinii* subsp. *suksdorfii* Caerulea Group) **(4)** contrast with foliage plants such as Bowles' golden sedge (*Carex elata* 'Aurea') **(5)**.

> **TOP TIP** YELLOW-VARIEGATED OR GOLDEN-LEAVED PLANTS, INCLUDING GRASSES SUCH AS *HAKONECHLOA MACRA* 'ALL GOLD' OR 'AUREOLA', ARE AN EXCELLENT CHOICE FOR USING AROUND SPRING BULBS TO BRIGHTEN SCHEMES WITH A FLASH OF BRILLIANT COLOUR, AS WELL AS TO COVER THE BARE SOIL AROUND THEIR BASE.

BLOSSOM BUDDIES

Spring-flowering trees and shrubs make the perfect partners for spring bulbs – there's nothing more jaw-dropping than a gorgeous garden scene of beautiful blossom above drifts of dreamy blooms. Top trees for blossom include Japanese flowering cherries, apple or crab apple trees, snowy mespil, magnolias, or hawthorn. Shrubs that flower at this time of year are often scented too, such as lilac, daphne, deutzia, abelia, and choisya. Add pretty self-seeders such as forget-me-nots around the bulbs to complete the picture.

RECREATE IT Under a lilac shrub (*Syringa vulgaris*) **(1)** and a stunning Japanese crab apple tree (*Malus × floribunda*) **(2)**, frothy forget-me-nots (*Myosotis sylvatica*) **(3)** and perennial honesty (*Lunaria rediviva*) **(4)** mingle with a sumptuous colour mix of tulips, including *Tulipa* 'White Dream' and purple 'Attila' **(5)**.

STARS OF THE SHADE

Woodland schemes really come into their own in mid- to late spring, with a multitude of terrific bulbs lighting up the shade under the emerging canopy. You can make the biggest visual impact in these plantings by choosing only two or three plants – for example, pairing hellebores with dog's tooth violets, or bluebells with ferns – but using a large number of each, and intermingling rather than planting in blocks.

RECREATE IT This gorgeous scheme features white double primrose (*Primula* 'Dawn Ansell') **(1)** with a mix of lily of the valley (*Convallaria majalis*) **(2)** and blue-flowered, silver-leaved Siberian bugloss (*Brunnera macrophylla* 'Jack Frost') **(3)**.

LATE SPRING TO MIDSUMMER

As the days get longer and warmer and spring turns into summer, a host of brilliant bulbs take advantage of the extra light and heat to bloom in abundance. This is the ideal time to create pretty cottage-garden schemes with ornamental onions and irises, or English country-garden style borders bursting with peonies and lilies. You can also bring the outside in by growing flowers such as anemones, ranunculus, and gladioli for cutting and making posies and bouquets. Alternatively, add a dash of drama with towering foxtail lilies and bold-coloured day lilies.

ALLIUM *ALLIUM*

These striking, globe-shaped flowers, borne on bare, straight stems, are the perfect plants for filling the flowering gap between spring and summer. The long-lasting blooms are usually purple but can be white, blue, pink, and even yellow, and are also beloved by bees.

BULB TYPE Bulb
HARDY Fully hardy
HEIGHT 0.2–1.8m (8in–6ft)
LEAF Long, strap-shaped; deciduous
POSITION Full sun, part shade

Allium hollandicum **'Purple Sensation'** has eye-catching, globe-like blooms that are adored by bees.

CHOOSE

Also known as ornamental or flowering onions, alliums come in hundreds of species and many more cultivars, in different flower colours and sizes, and to various heights. Most bloom in late spring and early summer, but there are some that flower as late as midsummer.

The most widely known and available alliums are the purple pompoms of *Allium hollandicum*, sometimes sold under the name *A. aflatunense*. Its cultivar 'Purple Sensation' grows to a height of around 90cm (36in). *Allium* 'Globemaster' is around the same height, but with larger flowerheads a little bit later, while *A. atropurpureum* has darker, reddish-purple flowers and more of a dome shape, and reaches 60cm (24in). The tallest types include *A.* 'Gladiator', which grows to 1.2m (4ft), and *A. giganteum*, giant allium, which reaches 1.8m (6ft), with huge spherical flowerheads up to 15cm (6in) across.

There is a range of white alliums from the species *A. nigrum*, black garlic, and *A. stipitatum* 'Mount Everest' to around 90cm (36in). *A.* 'Silver Spring' has pink dots in the centre of white florets. *A. karataviense* 'Ivory Queen' is just 20cm (8in) high, but still packs a punch with its pale, round blooms and wide leaves. *A. caeruleum* has blue flowers, and *A. moly* blooms yellow.

Other notable species include *A. cristophii*, star of Persia, which bears large silvery-purple globes that become more upright after blooming and hold colour and form well for many weeks. *A. schubertii* is a real showstopper, producing an astonishing firework

A. nigrum is one of the white alliums; its large flowers have distinctive green eyes.

'Silver Spring' has attractive, star-shaped florets with pink centres.

A. caeruleum has small, bright blue summer flowers.

of a flower. *A. siculum*, honey garlic, until recently sold as *Nectaroscordum*, looks completely different, with nodding, bell-like blooms in pale reddish-pink tinged cream.

A. sphaerocephalon, drumstick allium or round-headed leek, flowers later than the others, in midsummer, with small burgundy-hued blooms.

PLANT AND CARE

Alliums will tolerate part shade, especially honey garlic, but most of them prefer a sunny, sheltered spot. They don't like exposed sites that are cold and windy, or soggy soil.

Plant bulbs in early to mid-autumn in any well-drained soil. Place them at a depth that's four times the width of the bulb. Small types can be spaced about 8cm (3in) apart, and medium-sized ones around 20cm (8in), but larger bulbs should be separated by at least 30cm (12in) – this is in part to make sure their big flowerheads have the room they need. If growing in pots, make sure you plant them deep in compost with grit added to improve drainage. Some taller cultivars might need staking to help support the top-heavy blooms.

Allow the foliage to die back on its own before removing. On many types it will have yellowed by the time the flowers appear. Plant low-growing and

leafy plants around alliums in borders if you want to hide the leaves. Deadhead blooms after flowering if you wish, or simply enjoy the structural seedheads, which add a wonderful architectural element. Don't water the plants in the summer when dormant.

PROPAGATE Lift and divide or remove allium offsets in the late summer and replant them immediately.

The silvery-violet flowers of *A. cristophii* last for a long time.

WAYS TO GROW

Alliums are versatile and suit a range of situations, as long as the soil is well-drained. Many types have foliage that dies back early, making them great for naturalizing in drifts in grass, and they look wonderful woven through the middle of a border in groups. They work well in pots too, if the yellowing foliage can be disguised by surrounding plants. Later-flowering drumstick alliums look great with ornamental grasses in a prairie-style scheme.

You can also harvest ripe seed after flowering from most alliums and sow it into pots of gritty compost to grow on. However, if collecting seed from named cultivars, be aware that the resulting seedlings will probably not be the same as the original variety. Honey garlic has a reputation for self-seeding vigorously.

A. sphaerocephalon is a later-flowering allium that turns from green to burgundy.

WINDFLOWER *ANEMONE*

Anemones, or windflowers, have cup- or bowl-shaped poppy-like blooms and pretty fern-like leaves. A firm favourite with florists, they have silky petals in vibrant red, blue-violet, or white, and dramatic black centres. They're ideal for bridging the late spring gap in the garden.

BULB TYPE Tuber, rhizome
HARDY Frost hardy, fully hardy
HEIGHT 15–30cm (6–12in)
LEAF Deeply divided; deciduous
POSITION Full sun, part shade
WARNING! Toxic to pets. Ingestion may cause mild stomach upset. Contact can irritate the skin. Wear gloves and wash hands after contact.

Plant *Anemone coronaria* through the seasons for flowers in just a few weeks.

CHOOSE

Anemone coronaria have flamboyant, attention-grabbing blooms. The De Caen Group hybrids, such as deep-blue 'Mister Fokker' and scarlet-red 'Hollandia', have single flowers, while the frilly Saint Bridgid Group types are semi-double and double, and are usually sold in a mix of pink, blue, white, and red.

Windflower also refers to the fully hardy wood anemone – *A. nemorosa* has white, star-shaped blooms that appear in late winter and early spring. *A. blanda*, Grecian windflower, with daisy-like flowers, is another fully hardy spring beauty. Top cultivars include 'White Splendour' and 'Blue Shades'.

PLANT AND CARE

A. coronaria likes a light, well-drained soil in a sunny spot or in light shade. The tubers can be planted in different seasons throughout the year, for flowering 10 to 12 weeks later. However, they're sensitive to frost, so for best results, plant in early spring for flowers in early and midsummer. You can plant windflowers during the summer for autumn flowers. Alternatively, plant them in autumn for spring flowers, if you live in a mild area or have a greenhouse or conservatory where they can grow protected from frost.

Before planting, soak the tubers in warm to tepid water overnight. Plant them with the "claws" facing up, 5cm (2in) deep, about 10cm (4in) apart.

***A. blanda* 'White Splendour'** likes moist and shady woodland conditions.

WAYS TO GROW

Windflowers add bright colour to borders and pots in the late spring lull, and are especially valued as a cut flower. Wood anemones can be used to naturalize around the base of trees and shrubs, and as groundcover plants for shady banks or slopes.

After flowering but before the first frosts, lift the tubers, clean them, and keep in a cool, dark place to dry while dormant. In mild areas they may be left in the soil over winter, but should be mulched in autumn with organic matter.

Wood anemones prefer a lightly shaded position in moist but well-drained, fertile soil. Source freshly lifted rhizomes to plant in autumn. Soak in water overnight and plant them about 8cm (3in) deep, on their longest side, spaced at a minimum of 10cm (4in) apart.

Leave foliage of all types to die back on its own after flowering, before removing leaves or lifting. All types like to stay dry while dormant, so should not be watered during this period.

PROPAGATE *Anemone coronaria* can be divided when being lifted to dry in the dormant season, or the seed can be collected and sown. *A. nemorosa* spreads slowly via rhizomes. *A. blanda* will self-seed around on its own.

FUMEWORT *CORYDALIS*

BULB TYPE Tuber, rhizome
HARDY Fully hardy
HEIGHT 15–60cm (6–24in)
LEAF Divided; deciduous. Spreads to 30cm (12in).
POSITION Part shade, full shade

Fumewort, also known as bird-in-a-bush, has extraordinary blooms that comprise an inner tube of two petals and an outer spur-tipped pair. The flowers appear over a long period, from spring into early summer, and hang daintily above small mounds of intricately traced, fern-like foliage.

CHOOSE

Fumeworts can be annuals, biennials, or perennials that are evergreen or deciduous, with tuberous, rhizomatous, or fibrous roots, and different seasonal peaks. The best-known have blue flowers in spring and are deciduous, summer-dormant perennials.

Those grown from rhizomes include *Corydalis flexuosa* 'China Blue', with clusters of light-blue blooms to around 30cm (12in). 'Craigton Blue' has larger, bright-blue flowers on red stems, while *Corydalis* 'Tory MP' is the tallest, to 60cm (24in), with sweet-scented, brilliant-blue blooms and red-flushed stems. It sometimes reappears and flowers again in autumn.

Those fumewort that grow from tubers include pink-flowered *C. solida*, which is short and sweet at just 25cm (10in). The cultivars *C. solida* subsp. *solida* 'George Baker', 'Beth Evans', and 'Dieter Schacht' range from red to mauve-red to light-pink blooms.

PLANT AND MAINTAIN

Fumewort likes to grow in a cool spot in the shade, but will tolerate some sun if it stays moist. Plant the tubers or rhizomes in autumn, in fertile, humus-rich, moist but well-drained soil that's neutral to acid. Place them around 8cm (3in) deep and at least 20cm (8in) apart. After flowering, allow the foliage to die back on its own.

These plants are sometimes considered difficult, as they may not reliably return after their first season. However, this is mostly due to growing conditions, particularly their need to stay moist during dormancy in the summer. When planted in the ground, it can be easy to forget exactly where they are once their leaves have died back. If you grow them in pots, it can be easier to remember to water them at the same time as other containers.

PROPAGATE Every few years, after flowering, lift the tubers or rhizomes, carefully divide them (see p.27), and replant. Division helps keep fumewort coming back every year, so it's worth doing to ease congestion, even if propagation isn't the aim.

Corydalis flexuosa **'China Blue'** has sky-blue flowers and delicate leaves.

C. solida subsp. *solida* **'George Baker'** is low-growing with red petals.

ORCID *DACTYLORHIZA*

There are thousands of species of orchid, but only a few types suitable for growing in the garden. The genus *Dactylorhiza*, often called marsh and spotted orchids, are hardy, easy-to-grow perennials with small flower plumes in pink, purple, or white, sometimes with dotted petals and spotty leaves.

BULB TYPE Tuber
HARDY Fully hardy
HEIGHT 45–80cm (18–32in)
LEAF Lance-shaped; deciduous
POSITION Part shade, full sun

Dactylorhiza fuchsii is a great choice for meadow schemes and naturalizing in grass.

CHOOSE

Orchids are under threat in the wild, so always make sure to source your plants or tubers from a reputable supplier, such as a nursery that produces them in a laboratory

setting. Both plants and tubers are widely available, and are usually sent out when dormant, between the autumn and early spring, in a damp package.

Dactylorhiza fuchsii, the common spotted orchid, has densely packed white to pink flower spikes that grow to 60cm (24in), with darker pink spotting or markings on the petals. Although a wild flower, this species works well in the garden and also for naturalizing in long grass, and is a good orchid for beginners. It's more versatile than other orchids as it can grow in a wider range of situations and soils, and it sends up numerous flowers per plant.

D. purpurella, the northern marsh orchid, has been crossed with *D. foliosa* to create *D.* Foliorella gx, a hybrid with strong growth and deep purple blooms that can reach up to 80cm

Plant hardy orchid tubers in fertile, friable soil in spring or autumn.

WAYS TO GROW

Hardy orchids are most at home growing en masse in long grass and wildflower meadows, but they also look great in rock gardens and at the front of borders. They like damp conditions and are therefore perfect for adding a splash of colour on the banks of ponds.

(32in). *D. elata*, the robust marsh orchid, is even taller, at up to 90cm (36in), with reddish-purple flowers.

PLANT AND CARE

In general, orchids prefer part shade, but some will tolerate full sun if they stay damp. Plant in autumn or spring, in humus-rich, moist but well-drained soil. Place the tuber 2–3cm (1in) below the surface of the soil, and water well.

It's advisable to grow orchids on in large pots for their first year before planting out. Keep pots moist, watering with tap water if possible, and protect from frost in winter and early spring.

Allow foliage to yellow and wither completely before removing, to prevent disease. The flower stalks can be cut back to 5cm (2in) in autumn.

PROPAGATE Orchids will self-seed if happy and, once established, large clumps can be lifted and divided in autumn or early spring.

FOXTAIL LILY *EREMURUS*

Foxtail lilies grab all the attention with their statuesque blooms on straight, bare stems. The dramatic flower spikes are made up of hundreds of tiny, star-shaped blooms – magnets for bees and butterflies. Also known as desert candles, these plants can be fussy but are worth the trouble.

BULB TYPE Tuber
HARDY Fully hardy
HEIGHT 1–3m (3–10ft)
LEAF Lance or strap-shaped; deciduous. Spreads to 1m (3ft).
POSITION Full sun

CHOOSE

Popular, garden-worthy foxtail lily cultivars, such as the Ruiter hybrids, are often sold in a mix of colours, and include peach-hued *Eremurus* 'Romance' and *E.* 'White Beauty Favourite', with its white spikes. *Eremurus × isabellinus* 'Cleopatra', with orange-pink blooms, is widely available and grows to 1.5m (5ft) tall. Around the same height is *E. stenophyllus*, with yellow flowers. Larger forms, with tall stems and wider spread, include white *E. himalaicus*, to 2.5m (8ft), and pale-pink *E. robustus*, which reaches a whopping 3m (10ft) high and 1m (3ft) wide.

Soft pink flower spikes of *Eremurus robustus* tower atop tall, straight stems.

PLANT AND CARE

Foxtail lilies are particular about where they grow. They need a sunny spot, sheltered from strong winds. These plants don't do well where the young foliage can get damaged by frost, and they hate wet soil and being disturbed or crowded by other plants. You'll get best results in a fertile, sandy, very well-drained soil that's neutral to acidic.

Foxtail lilies are available as ready-to-grow pot plants, but you can source bare-root tubers in autumn or spring: in autumn, the tubers are more likely to be fresh. Often compared to a starfish, they have a central crown with fleshy roots that radiate outwards. Soak in tepid water overnight and plant as soon as possible after receipt, handling gently.

Dig a hole around 15cm (6in) deep and 30cm (12in) across or wide enough to hold all the roots. Add grit or sand to improve the drainage, if necessary. Create a small mound of soil or grit in the centre of the hole. Place the crown on top, with the fleshy roots fanned out around it. Cover the roots with soil, but leave the crown at or just under the surface. Space the plants at least 30cm (12in) apart. Water in well.

Feed foxtail lilies during the growing season with a general liquid fertilizer. The flower spikes may need staking in exposed gardens, but be careful not to damage the roots while inserting the support. The flower stalk can be removed after blooming (unless you want it to self-seed), but let the leaves

WAYS TO GROW

Use foxtail lilies to create height at the backs of borders or as striking vertical accents in the middle of planting beds. They're not a good choice for pots because of their wide-spreading foliage but work well in prairie-style schemes and dry or gravel gardens.

The stately *E. stenophyllus* has spectacular, golden-yellow blooms.

wither before cutting them back. Mulch in autumn, but leave the crown uncovered. They are perennial plants, but may not return in unsuitable conditions, such as waterlogged soil or if the roots are damaged.

PROPAGATE Foxtail lilies will self-seed, so don't deadhead the flowers until late summer or autumn if you want them to spread. They can also be lifted and divided after flowering every few years with a fork – take extreme care not to damage the large but delicate roots.

GLADIOLI *GLADIOLUS*

Gladioli, also known as corn flag, have tall, stately flower spires packed with masses of funnel-shaped blooms, in bright colours from pink, purple, and white to yellow, red, and even green. They open in sequence, from bottom to top, and are great for pollinators.

BULB TYPE Corm
HARDY Half-hardy, tender
HEIGHT Up to 1.5m (5ft)
LEAF Linear, sword-shaped; deciduous
POSITION Full sun

CHOOSE

The classic, large-flowered gladioli, often grown for cut flowers, are around 90cm–1.2m (36in–4ft) tall. There are many cultivars in every hue, and bi-coloured types. Popular plants include lime-green G. 'Evergreen', red 'Zorro', white 'Bangladesh', and cream 'Sancerre'. The smaller Butterfly hybrids are better than large-flowering types as garden plants in a bed or border.

Gladiolus nanus varieties, also known as dwarf gladioli, are shorter and smaller again, to 70cm (27in), and somewhat hardier too. Their star-shaped or orchid-like blooms are good for growing in pots.

Large-flowered gladioli, like creamy-white 'Sancerre', make lovely cut flowers.

G. 'Prins Claus' is a fine choice, with white flowers with pink and red splashes. *G. communis* subsp. *byzantinus* has slender, graceful spikes that are less densely packed, in magenta-pink. It grows to 90cm (36in) and will self-seed and naturalize when happy.

PLANT AND CARE

Gladioli like a sheltered spot in full sun. They need light, humus-rich, moist but well-drained soil, and will rot in wet conditions or heavy soils.

Plant the corms in spring, at least 15cm (6in) deep, and 15–20cm (6–8in) apart, with the pointed end facing up. Water well and keep soil moist during the growing season. Plant a group of corms every week or two, so you have flowers over a long period, or all at once for a stunning one-off show. In cold areas with late frosts, during wet springs, or on heavy clay soil, start them off in pots somewhere frost-free, and transfer into the ground in late spring. Alternatively, wait and plant the corms out in mid- to late spring after all risk of frost has passed.

Feed with tomato fertilizer in the growing season, and stake taller and larger-flowered types as they grow, avoiding damage to the underground corm when inserting the cane. Remove individual spent flowers on the stem as they fade. Gladioli can be attacked by thrips, slugs, and aphids. The corms need to stay frost-free, so should be lifted to overwinter indoors (see p.22).

PROPAGATE Congested clumps of gladioli may cease blooming, so be sure to divide them every few years. Remove offsets when lifting the plants during the autumn.

WAYS TO GROW

Gladioli are the absolute queens of the cottage garden. They add height if drifted through summer borders and are a staple of the cut-flower patch. Those with hot colours are magnificent for schemes with a tropical look. *Gladiolus communis* subsp. *byzantinus* naturalizes well in grass.

G. communis subsp. byzantinus is great in summer meadow plantings.

DAY LILY *HEMEROCALLIS*

Day lilies are easy to grow in almost every soil or situation. As their name suggests, the large funnel-shaped blooms – in warm hues of orange, red, pink, cream, and yellow – last for just one day, but are produced in abundance over a long period through to early autumn.

BULB TYPE Tuberous rhizome
HARDY Fully hardy
HEIGHT Up to 1.2m (4ft)
LEAF Grass-like; evergreen or deciduous. Spread to 1.2m (4ft).
POSITION Full sun, part shade

Hemerocallis **'Stafford'** is one of the best red cultivars of day lily.

H. **'Custard Candy'** will grow well in almost any type of soil.

CHOOSE

The intensive breeding of day lilies has resulted in a huge array of cultivars, which are available in many heights, flower forms, sizes, colours, and blooming times, from early to late season. The blooms can be trumpet- or star-shaped, double, ruffled, or recurved, with broad or narrow petals.

Hemerocallis lilioasphodelus, previously known as *H. flava*, is one of the earliest to bloom. It has fragrant, sunny yellow, lily-like flowers and reaches 50cm (20in). *H.* 'Stafford', which reaches 70cm (28in), has dark red flowers with a yellow centre and much narrower petals. *H.* 'Custard Candy' is a big,

trumpet-flowered type that's highly floriferous, with cream petals and a maroon band around the yellow-green centre. *H.* 'Catherine Woodbery' has pretty, mauve-purple to pink-washed flowers with a yellow centre.

PLANT AND CARE

Day lilies are tough and adaptable, and come back reliably every year. They settle in quickly and are long-lived and almost trouble-free, requiring little care and attention. These plants prefer full sun but will still grow if they're in shade for several hours a day. Their ideal soil is fertile, moist, and well-drained, but they'll also tolerate heavy soils

such as clay, and poor soils too. Once they're established, day lilies will also cope with drought.

Plant bare-root day lilies in autumn or spring. Dig a hole that's large enough to accommodate the crown and roots comfortably. Cover with soil and gently firm with your hands, making sure the crown section is just under the surface. Water well and keep moist during dry periods in their first season.

Deadhead the spent flowers as they fade and remove the flower stem when it has finished blooming. Let the foliage on deciduous types yellow and die back on its own before removing.

Watch out for damage from slugs and snails (*see p.29*). Early-flowering types can suffer from gall midge.

PROPAGATE Divide established clumps every three to four years, after flowering, in late summer or early autumn – this will also keep plants vigorous and blooming freely.

WAYS TO GROW

Day lilies will grow virtually anywhere, but are particularly good for slopes and banks, groundcover, and low-maintenance prairie-style schemes, as well as dry and gravel gardens. They look great in borders and containers, as long as the pot is large enough and refreshed with new compost every year.

IRIS *IRIS*

Intricate and elegant, iris flowers have a real wow factor. They come in a huge variety of forms and colours, but mostly comprise three upright petals, called standards, above three downward arching petals, called falls. Petals can be veined or splashed with contrasting colours, or feature a hairy "beard".

BULB TYPE Bulb, rhizome
HARDY Fully hardy
HEIGHT 0.5–1.2m (1½–4ft)
LEAF Sword-shaped or grass-like; evergreen, deciduous. Spreads to 60cm (24in) or more.
POSITION Full sun, part shade
WARNING! Ingestion may cause severe discomfort. Wear gloves and wash hands after contact.

Iris **'Jane Phillips'** is a classic and one of the most well-loved bearded irises.

I. *pseudacorus*, yellow flag, is great near water but will overwhelm small ponds.

'Kent Pride' is rich copper-brown with a golden central "beard".

CHOOSE

There are hundreds of species of irises, but the most popular late-spring and summer-flowering types for gardens are evergreen bearded and flag irises – which are grown from rhizomes – and deciduous Dutch irises, which grow from bulbs.

Iris germanica, bearded irises, reach around 90cm (36in) tall, and have unashamedly big, frilly, blowsy blooms, with ruffled petals and fuzzy hairs along the centre of the falls. As well as the flowers, they offer another interesting feature with their glaucous,

upright, lance-shaped leaves. Two of the best are pale-blue 'Jane Phillips' and dark copper and burgundy 'Kent Pride'. Almost as showy is another bearded type, *I. pallida* subsp. *pallida*, known as orris root or sweet iris, which has lovely lavender-blue colouring.

I. sibirica, Siberian flag iris, is typically smaller and looks more delicate. 'Perry's Blue' has clear blue veined flowers with white marks, while 'Tropic Night' has violet flowers with yellow marks.

Iris × *hollandica*, known as Dutch iris, grow from bulbs to about 60cm (24in) in height. They have strap-like, deciduous leaves and are great for

cut flowers. Some of the most widely grown include 'Apollo', with yellow and white blooms; 'Tigereye', with purple and brown-red flowers with yellow marks; and 'Red Ember', which has bronze-red and purple petals.

I. pseudacorus, yellow flag, has bright yellow flowers and reaches over 90cm (36in). It's known for growing beside and in water, but can become invasive. The cultivar 'Crème de la Crème' is a paler cream colour. *Iris* × *robusta* 'Gerald Darby' is a pretty, floriferous cultivar with purple and white petals. It reaches 90cm (36in) and tolerates some shade.

Siberian flag irises such as 'Perry's Blue' have beautiful, intricate veining and markings on the fall petals.

PLANT AND CARE

Bearded irises thrive in a sheltered and preferably south-facing situation in full sun, because they need to bake in its rays for most of the day during the summer to bloom well the following season. Plant the bare-root rhizome in autumn or spring, in fertile, well-drained soil. Soak in tepid water for up to an hour before planting. Plant so that the crown sits just above the soil. Water in well. Deadhead spent flowers. Don't let bearded irises get crowded by other plants or weeds. Cut back the leaves during the autumn to prevent wind-rock (when the wind rocks the plant out of the soil). Keep a look out for fungal diseases such as rhizome rot (see p.29) and iris leaf spot, removing any affected material.

Siberian flag iris prefers a moist soil in sun or part shade, and thrives as a marginal plant beside water. Plant in late spring to late summer, positioning the rhizome just above the surface. Place at least 30cm (12in) apart to accommodate its vigorous clumping habit.

With bulbous irises such as Dutch iris, plant the bulbs in autumn, in a fertile, well-drained soil in full sun. Plant them 10cm (4in) deep and 10cm (4in) apart. Allow the foliage to die back on its

I. x hollandica 'Apollo' needs well-drained soil and sunshine to perform.

WAYS TO GROW

Bearded irises associate beautifully with other plants in borders but are often grown as a feature in their own bed. Siberian and flag irises work well around ponds and in bog gardens. Dutch irises also complement any late spring and early summer scheme and are popular cut flowers.

own before removing any leaves. Dutch irises can be perennial, but are frequently treated as annuals with new bulbs planted every year to ensure a reliable display.

PROPAGATE Divide rhizomatous irises after flowering, during midsummer, splitting and replanting them straight away (see p.27). Established clumps will get congested and then require thinning or dividing every three years. Divide your Dutch irises in autumn, lifting and removing offsets to pot up and grow on.

Dutch iris 'Lion King' has exquisite reddish standards and golden, striped falls.

LILY *LILIUM*

Big, bold, and showy, lilies are garden classics adored for their large trumpet-shaped flowers atop tall, elegant stems – the perfect cut flower. Easy to grow and reliably perennial, they come in colours from pink, purple, and white to yellow, orange, and red, both fragrant and unscented.

BULB TYPE Bulb
HARDY Fully hardy
HEIGHT 0.6–1.5m (2–5ft)
LEAF Lance-shaped or whorled; deciduous
POSITION Full sun, part shade
WARNING! All parts are toxic to pets, especially cats

Lilium **'Grand Cru'** are easy-care, trouble-free plants.

CHOOSE

Lilies vary wildly in size, with an array of flower shapes, colours, and growing needs. There are thousands of cultivars to choose from. The most popular, widely available types are divided into groups, including Asiatic, Oriental, and Longiflorum Asiatic. Asiatic lilies, such as yellow 'Grand Cru', are the easiest to grow, earliest to bloom, and some of the shortest, growing to just 90cm (36in), but they aren't scented. Oriental lilies, such as white 'Casa Blanca' and dark pink, upward-facing 'Star Gazer', are heavily fragrant, flower mid-season, and reach about

1.2m (4ft). The Longiflorum Asiatic types, including 'Eyeliner', with its cream flowers and black-edged petals, combine scent and ease of growing.

Lilium regale, trumpet lilies, have huge, funnel-shaped, white blooms to 15cm (6in) across, with yellow centres and dark pink, flushed reverse on the petals, and strong scent. Top cultivars include 'Pink Perfection' and orange 'African Queen'.

Reliable and tough, *L. martagon*, Turk's cap lilies, produce an abundance of small, downward-facing, pendent blooms with rolled back or reflexed petals in various shades of purple.

'Casa Blanca' is one of the best white-flowered Oriental lilies.

Good varieties include the purple-red 'Claude Shride'. Similar-looking but later-flowering is *L. speciosum* var. *rubrum*, which is deep pink and reaches 1.2m (4ft). *L. henryi*, another floriferous Turk's-cap type, with up to 20 orange blooms per stem, reaches 1.5m (5ft).

PLANT AND CARE

In general, lilies love full sun, but with their roots kept cool, perhaps using the light shade that's created by the foliage of other plants. Some, such as *Lilium martagon* and *L. speciosum*

L. speciosum var. *rubrum* has elegant recurved petals with white edges.

The upward-facing blooms of 'Star Gazer' have a strong, heady scent.

Large, showy lilies such as Asiatic and Oriental varieties look terrific in beds and borders, as well as in large deep containers for a stunning seasonal display. *Lilium speciosum* var. *rubrum* and *L. martagon* are more at home in the shade of a woodland scheme, where they'll slowly naturalize over time.

Turk's cap lily 'Claude Shride' grows best in shady, informal areas.

var. *rubrum*, prefer part shade. Lily bulbs can be planted in autumn or spring – spring is the best choice for gardens with heavy soils or wet winters, and this is often when you'll find the bulbs for sale. Most lilies

L. henryi has multiple small, copper-hued flowers dangling prettily from each stalk.

need fertile, well- drained soil; you can improve the planting hole with organic matter, if necessary. *L. speciosum* var. *rubrum* prefers it moist but well-drained and, like Oriental lilies, enjoys acidic soil. Grow them in ericaceous compost in pots if the ground conditions aren't right. The Longiflorum Asiatic hybrids and *L. regale* are the least fussy, and will grow in most soil types.

Depending on their size, plant the bulbs at least 20–30cm (8–12in) deep and 30cm (12in) apart. In wet areas, plant them on their sides so they don't rot.

Keep plants well-watered and feed with liquid fertilizer during the growing season. Taller types may need staking. Deadhead the flowers as they fade to encourage more blooms, but leave the stems up until autumn and allow foliage to yellow and die back on its own before removing. Mulch with organic matter in spring. If growing in a pot, you can happily leave them in the same pot for several years.

Keep a look out for attack from lily beetles (*see p.28*) and aphids, removing them from the plant immediately.

PROPAGATE Every few years, clumps of lilies can be lifted and divided after flowering, once the foliage has died back. Offsets, including bulblets and stem and aerial bulbils, can be removed at the same time to grow on. Alternatively, you can propagate lilies by scaling. Once the foliage has died back, lift the plant and take scales off the sides of the bulb (*see p.25*). You can also allow the plants to form seedpods, instead of deadheading the flowers, and collect fresh seed in late summer. Sow the seeds into pots of compost to grow on somewhere frost-free. If successful, both these methods of propagation can take from three to five years to produce a flowering plant.

STAR OF BETHLEHEM

ORNITHOGALUM

Also known as chincherinchee and wonder flowers, these rare beauties vary in size and hardiness, but all feature eye-catching, densely packed clusters of star-shaped blooms in white, yellow, or orange. Their showy looks and upright stems make them excellent long-lasting cut flowers.

BULB TYPE Bulb
HARDY Tender, half-hardy, fully hardy
HEIGHT Up to 90cm (36in)
LEAF Strap-shaped; deciduous
POSITION Full sun, part shade
WARNING! Ingestion may cause severe discomfort. Sap can irritate the skin. Wear gloves and other protective equipment; wash hands after contact.

CHOOSE

Ornithogalum thyrsoides, wonder flower, is one of the most popular plants in this genus. It reaches up to 60cm (24in) and has conical flower spikes packed with up to 30 cup-shaped, white flowers with green centres.

O. saundersiae, giant chincherinchee, is a real showstopper, with flatter flowerheads of multiple creamy-white blooms with black centres. It grows to around 90cm (36in) high. *O. dubium* has similar flowers, but in striking, bright orange or yellow, and reaches 30cm (12in). These three species are tender or half-hardy and won't survive a cold winter outside.

O. nutans, drooping star of Bethlehem or silver bells, is hardier and may be able to survive winter outdoors. It grows to 45cm (18in), and has pretty, more loosely arranged flowerheads of up to 15 downward-facing white blooms. Broad grey-green stripes mark the back of each petal.

O. umbellatum, known as common star of Bethlehem, is fully hardy and spreads vigorously. It reaches 15–30cm (6–12in).

PLANT AND CARE

Star of Bethlehem likes a warm spot in a sheltered west- or south-facing position. It will tolerate light shade

WAYS TO GROW

Tender and half-hardy bulbs are often planted in containers for a stunning seasonal display and ease of lifting afterwards. Hardier types that are left out all year are perfect for naturalizing on sunny banks.

but grows and flowers best in sun, and also prefers fertile, moist but well-drained soil. Plant half-hardy and tender types in spring. Wait until the risk of frost has passed before planting in the ground, or pot up and grow on in a frost-free spot to plant out once the risk of frost has passed. Plant hardier *O. umbellatum* and *O. nutans* out in autumn if leaving in the ground year-round.

Plant bulbs 5–7cm (2–3¾in) deep, 13–15cm (5–6in) apart, pointed ends facing up. Water well in the growing season but keep dry while dormant. After flowering, when the leaves have died back, lift and store tender and half-hardy bulbs somewhere cool and frost-free. Alternatively, grow them in pots and move indoors over winter, and out into the garden again in spring.

PROPAGATE Remove offsets when lifting bulbs or while they're dormant, and pot up to grow on. Harvest seed when ripe and sow fresh.

Ornithogalum thyrsoides holds masses of white flowers on tall, elegant stems.

Small but striking, O. dubium is perfect for containers.

WOOD SORREL *OXALIS*

These pretty, low-growing perennial plants have green or purple shamrock-shaped leaves, which often droop and fold up at night. Most have pink or white flowers that open in sunshine. Some are seen as invasive weeds, but several are considered excellent ornamental bulbs for the garden.

BULB TYPE Bulb, tuber
HARDY Tender, half-hardy
HEIGHT Up to 30cm (12in)
LEAF Three-part, clover-like; deciduous. Spreads to 30cm (12in).
POSITION Part shade, full sun

Oxalis triangularis subs. *papilionacea* is a prolific flowerer with loads of blooms.

CHOOSE

Oxalis triangularis subsp. *papilionacea*, purpleleaf false shamrock, is a bulbous plant that produces striking, deep purple clover-like foliage from spring to autumn and white or pale pink star-shaped flowers in summer. It grows into a mound about 20cm (8in) high and 30cm (12in) wide.

O. tetraphylla 'Iron Cross', the good luck plant, is a tuber with green leaves that have maroon-purple splashes at the centre, and deep coral-pink flowers. It reaches about 30cm (12in) high and 15cm (6in) wide.

PLANT AND CARE

O. triangularis isn't fully hardy and can be grown outdoors if lifted and stored over winter, or in a pot that can be placed out in summer and brought in for winter. 'Iron Cross' is tougher and may survive outside in mild gardens, but should be lifted and stored over winter in cold or wet climates. Both appreciate a sheltered position in part shade or full sun. Plant out in spring once all risk of

O. tetraphylla '**Iron Cross**' has pink blooms and burgundy-splotched leaves.

frost has passed. Alternatively, start off in pots and transplant out into the ground once all risk of frost has passed.

Plant wood sorrel in fertile, moist but well-drained soil. Place bulbs about 5cm (2in) deep and at least 15cm (6in) apart; and place tubers about 2cm (¾in) under the surface of the soil. Water well and keep moist during the growing season. Remove the withered foliage once it has died back, in late autumn.

PROPAGATE Divide large or congested clumps of this plant while dormant. Many wood sorrels spread easily through self-seeding.

WAYS TO GROW

Wood sorrels work well as groundcover and edging to beds and borders, as well as in rock gardens. Often they're grown indoors as houseplants in pots on windowsills or in a conservatory or greenhouse, and moved outside for summer.

Wood sorrels are excellent plants for edging beds and low groundcover.

PEONY *PAEONIA*

Peonies light up the border with their magnificent, blowsy blooms in early summer. They grow slowly and only flower for a short time, but they're extremely long-lived plants, and well worth waiting for, with their big, bowl-shaped fragrant flowers in shades of red, pink, white, and yellow.

BULB TYPE Tuber
HARDY Fully hardy
HEIGHT 60cm–90cm (24in–36in)
LEAF Divided; deciduous. Spreads up to 60cm.
POSITION Full sun, part shade
WARNING! Ingestion can cause stomach upset. Wear gloves and wash hands after contact.

CHOOSE

Among the most popular herbaceous peonies are *Paeonia lactiflora* 'Duchesse de Nemours', with its white, fragrant flowers, like scrunched-up tissue paper, with pale-yellow centres, and 'Bowl of Beauty', which has silky pink outer petals around a crinkled, cream centre. Both plants reach up to 90cm (36in).

A little later to flower is *P. lactiflora* 'Sarah Bernhardt', with huge, double, ruffled rose-pink blooms. It's beautifully scented, makes an excellent cut flower, and reaches around 90cm (36in).

P. daurica subsp. *mlokosewitschii*, known as Molly the Witch peony, is smaller, with blue-grey leaves and lovely, single, pale yellow blooms.

'Bowl of Beauty' has pink outer petals, which frame stunning centres with masses of ruffled, cream petals.

Another group of peonies that grow from tubers are known as Itoh or intersectional peonies, produced by breeding a cross between herbaceous peonies and tree peony shrubs (the latter don't grow from tubers). These plants have very large flowers in a range of forms and colour combinations. Unlike the herbaceous types, they may not need staking and can be grown in large containers. The foliage may also last longer and turn different colours in autumn, and they tend to be more compact in growth, reaching about 75–90cm (30–36in) in height. These are highly prized plants and are often more expensive than other types. Widely available cultivars include yellow *P.* 'Bartzella', and *P.* 'Cora Louise', which has big, pale pink flowers with a large, purple-red splash at the centre.

PLANT AND CARE

Peonies like a sheltered position in full sun, or part shade with several hours of sun a day. They need a fertile, well-drained soil that's preferably neutral to alkaline. These plants hate sitting in wet soil, as well as being crowded by other plants as they develop large roots that need a deep root run. For best results, plant them at a distance from trees and shrubs.

Pot-grown plants are available, which can be planted in spring or autumn. If you're buying bare roots, plant them

'Duchesse de Nemours' is a classic peony, with large, double, white flowers.

'Sarah Bernhardt' is big and blowsy with a gorgeous, subtle scent.

P. daurica subsp. *mlokosewitschii* is the best yellow peony.

WAYS TO GROW

Herbaceous peonies are best planted in the ground. Itoh peonies can tolerate growing in a large pot for a period, but as peonies develop large root systems over time, their preferred home is in a bed or border. They look particularly at home in cottage and country schemes with other English-garden style perennials.

in autumn. Dig a large hole of around 30–40cm (12–16in) by 30cm (12in), and add organic matter to enrich the soil. Plant the tuber so the crown with buds or "eyes" at the top is no more than 2.5–5cm (1–2in) below the soil's surface: peonies won't flower if planted too deep. Water in; don't let it get too dry while establishing. Once settled, peonies are quite drought-tolerant.

These plants can sometimes take time to establish. To help them grow strong, don't cut any of the flowers for the vase in the first year. After this, make sure to leave at least a third of the blooms on each plant so they can build energy for future seasons.

Herbaceous types may need staking to support the heavy flowerheads. To get the largest blooms, remove the side-buds near the bottom of each main flower bud – simply cut them off in late spring. Leave them alone if you prefer to have more, albeit slightly smaller, flowers over a longer season.

Deadhead spent blooms. You can cut the withered foliage back to ground-level in autumn. If mulching beds or borders, be careful not to cover the crowns.

In spring, you may see ants swarming all over the flower buds, as they're attracted to the nectar they secrete. However, these insects won't do any harm and aren't a problem.

Watch out for damage from a fungal disease called peony wilt, which causes the stems and buds to turn brown and wither in spring. Immediately remove any infected material and burn or bin it, making sure not to compost it.

PROPAGATE Peonies have a reputation for not liking to be disturbed. The tubers can be lifted and divided (*see p.27*) in autumn, but each piece may then take time to re-establish.

The purple-centred Itoh peony 'Cora Louise' has showy blooms that last longer than other types.

TUBEROSE *POLIANTHES TUBEROSA*

Tuberose are beloved for their sensationally scented blooms. Each upright stem produces between 30 and 50 white, waxy flowers, which are borne in pairs. The exquisite, long-lasting blooms open in succession over a long period, and perfume the evening garden with their delicious fragrance.

BULB TYPE Tuber
HARDY Tender
HEIGHT Up to 1.2m (4ft)
LEAF Lance-shaped, deciduous
POSITION Full sun

The fragrant flowers of tuberose open from bottom to top.

CHOOSE

Polianthes tuberosa, also known as *Agave amica*, is originally from Mexico. It has different names throughout the world, including tuberose, polyanthus lily, and Hawaiian garland flower.

'The Pearl' is a tall and showy double-flowered cultivar.

P. tuberosa has single white flowers and grows to around 90cm (36in). The cultivar 'The Pearl' has double white flowers and is taller, to 1.2m (4ft). There's also a pink-flowered variety called 'Sensation', and one with yellow blooms called 'Super Gold'. All have a heady, sweet scent.

PLANT AND CARE

Tuberose will grow best in a warm spot in full sun, sheltered from drying winds. They prefer loamy, alkaline to neutral, well-drained soil, and although they need to stay moist while growing, they hate being waterlogged. Tubers should be planted around 8cm (3in) deep, spaced at least 15cm (6in) apart in the ground, and watered in well.

They can be planted straight out in spring in mild climates, once the temperature is consistently over about 13°C (55°F). However, where frosts continue through spring, start them off undercover in pots in early spring and then transplant outdoors once all risk of frost has passed.

Once the leaves emerge, keep the young plants moist and feed with liquid fertilizer every two weeks.

Tuberose are usually grown as annuals in the garden with new tubers planted each year, but you can try to keep them and grow them as a perennial – with the caveat that you may only get flowers every other year. Once the foliage has died back, remove the leaves and lift and store tubers undercover overwinter, or bring those grown outdoors in containers indoors in autumn.

PROPAGATE Offsets can be removed when the plant is lifted at the end of the growing season.

WAYS TO GROW

Grow tuberose where the scented flowers can be best enjoyed, in pots on the patio, or in beds and borders alongside seating areas. They're often grown under glass such as a greenhouse or conservatory, or as a houseplant, but are most popular as cut flowers.

PERSIAN BUTTERCUP

RANUNCULUS ASIATICUS

These unusual blooms have masses of crepe-paper-like petals, tightly packed in ruffled layers, and available in a rainbow of bright colours from red, orange, and pink, to yellow and white. They produce an abundance of long-lasting, large flowers that can reach 15cm (6in) wide.

BULB TYPE Corm
HARDY Half-hardy
HEIGHT Up to 45cm (18in)
LEAF Divided; deciduous
POSITION Full sun

Bright, long-lasting blooms make Persian buttercup a favourite cut flower.

CHOOSE

The Persian buttercup is also known as the rose of spring. The species has single, cup-shaped flowers with black centres on straight, branching stems above a mound of fresh, fern-like foliage. It's loved by florists, so many hybrids have been bred offering double or frilly blooms that are sometimes bi-coloured or with contrasting petal tips.

Most Persian buttercups are sold as an unnamed colour mix, but there are several cultivar series worth seeking out, such as the Tomer hybrids, with their peony-shaped flowers, the Tecolote varieties in shades of red and pink, the Aviv and Bloomingdale hybrids, and the Butterfly Series, which have mildew resistance.

PLANT AND CARE

In mild climates, plant corms out in the ground in autumn. In colder and more exposed gardens, plant them in pots to overwinter under glass, in an unheated greenhouse or conservatory, or on a bright sill somewhere cool. Keep well-ventilated. Plant out once all risk of frost has passed. Wait for spring to plant corms in the ground for summer flowers, or buy potted plants in spring.

Persian buttercups bloom best in full sun, or in part shade with several hours of sun a day, in a sheltered spot. Grow in fertile, moist but very well-drained, sandy or loamy, slightly acidic soil. Add grit to growing mix for container plants.

Soak the corms in tepid water for a few hours before planting. They have an "eye" or bud at the top, and what look

Plant Persian buttercup with the "claws" at the bottom facing down.

like claws at the bottom. Plant 5cm (2in) deep, spaced at least 10cm (4in) apart, with the "claws" facing down. Water well, but don't irrigate again until leaves emerge. Keep the plants moist during active growth, until the leaves yellow.

Persian buttercup can be damaged by the fungal disease powdery mildew. Affected material should be removed from the plant and destroyed, and tools and hands thoroughly cleaned afterwards to prevent spread.

Deadhead spent blooms to encourage more flowers, but don't remove the foliage until it has died back on its own. These plants like to stay dry while dormant and can be frost tender, so if you wish to treat them as perennials, they should then be lifted and stored under cover.

PROPAGATE Harvested seed can be sown while ripe but resulting seedlings probably won't match the original cultivars. Divide plants in autumn.

WAYS TO GROW

Persian buttercups make a colourful display in containers, and are great as seasonal fillers, used like bedding, in border schemes. They're a staple of the cut-flower garden and last well in arrangements.

HARLEQUIN FLOWER

SPARAXIS

As their name suggests, these starry, trumpet-shaped flowers are luminous in a bright mix of colours, from orange and white to plum, red, and yellow. Each small, wide-open bloom has a central throat in contrasting shades and intriguing graphic patterns.

BULB TYPE Corm
HARDY Tender
HEIGHT 25–45cm (10–18in)
LEAF Lance-shaped; deciduous
POSITION Full sun

Sparaxis tricolor have bold, contrasting colour markings at their centres.

CHOOSE

Sparaxis tricolor, harlequin flower or wandflower, bears short, wiry stems to around 45cm (18in), topped with a spike of loosely clustered, funnel-shaped flowers in a blazing mix of bright colours. The vibrant yellow centres are often framed by dark or black markings. The more demure *Sparaxis elegans*, Cape buttercup, 30cm (12in), carries up to five white, apricot-pink, or orange blooms per stem. Each has a purple central throat. *Sparaxis grandiflora* is even shorter, to just 25cm (10in), and produces clusters of rich purple flowers.

PLANT AND CARE

These tender flowers can only be grown outdoors all year in mild, frost-free climates. In colder areas they can be enjoyed indoors, or cultivated so they bloom in the garden in late spring or early summer. To do this, start off corms in pots under glass, in an unheated greenhouse, conservatory, or a similar bright, frost-free place, in late autumn. As the plants grow, keep the compost just moist but not wet. Transplant into the ground in spring, or move display containers outdoors, once all risk of frost has passed. Alternatively, for flowers later in summer, wait until risk of frost has passed and plant the corms outdoors in the ground.

Harlequin flowers prefer fertile, well-drained, sandy soil and a warm sheltered position in full sun. Plant corms about 15cm (6in) deep, spaced at least 10cm (4in) apart.

Deadhead spent flowers but leave foliage to die back on its own. Treat as an annual, using fresh corms each year, or as a perennial by lifting and storing the corms undercover while dormant.

PROPAGATE If you wish to harvest seed, don't deadhead spent flowers, but allow them to develop into seedheads. Collect and sow undercover. Remove offsets while the plants are dormant.

WAYS TO GROW

Harlequin flower is often grown as part of a glasshouse display or even as a houseplant. It makes a colourful summer bloom for outdoor pots, borders, and rock gardens, and for cutting.

S. grandiflora subsp. grandiflora is short but sweet in vivid purple.

ARUM/CALLA LILY *ZANTEDESCHIA*

Make a statement with these striking architectural blooms. Each flower comprises a central spike, or spadix, and a large hood-shaped, petal-like bract that wraps around it, called the spathe, and sits upon a fleshy, tubular stem above glossy, sometimes blotched, foliage.

BULB TYPE Rhizome
HARDY Fully hardy, half-hardy, tender
HEIGHT 40–90cm (16–36in)
LEAF Arrow-shaped; semi-evergreen
POSITION Full sun, part shade
WARNING! Ingestion may cause severe discomfort. Sap may irritate the skin and eyes. Wear gloves and other protective equipment. Wash hands after contact.

The arum lily *Zantedeschia aethiopica* 'Crowborough' looks delicate but is tough.

The calla lily *Zantedeschia* 'Picasso' has painterly purple-flushed bracts.

CHOOSE

There are hardy forms of *Zantedeschia*, such as *Z. aethiopica*, which are usually called arum lilies, and less hardy forms, such as *Z. elliottiana*, which are referred to as calla lilies. However, these terms are often used interchangeably.

One of the most popular and hardiest forms is *Z. aethiopica* 'Crowborough', which has striking pure-white blooms. Over time it will form large clumps. 'Green Goddess' has green-flushed spathes and is frost-hardy to around -5°C (23°F) so may survive winter.

Less hardy to tender cultivars include 'Lipstick', with bright pink blooms, 'Odessa', which is a dramatic dark purple-black; and bi-coloured 'Picasso', with cream to green spathes with violet-flushed interiors.

PLANT AND CARE

Zantedeschia need a sheltered position in full sun or part shade with several hours of light a day, in cool, fertile, moist soil. Plant hardy types in the ground in spring, and half-hardy and tender types in pots under cover,

transplanting them outside if you wish to do so only after all risk of frost has passed.

Plant shallowly, so that the top of the rhizome is just at the surface of the soil, and space them at least 45cm (18in) apart. Keep moist by watering through the growing season if in the ground or pots. Hardy types can also be grown in a pond, where the water is no more than 30cm (12in) deep, in a pond basket with aquatic compost.

Outdoor-grown plants will be evergreen in warm climates, and deciduous in colder ones (semi-evergreen). Remove the foliage once it has died back. Mulch those left outside over the winter in autumn to protect their roots. Move pots of less hardy types indoors, or lift the rhizomes and those grown in pond baskets, and store in a frost-free spot over winter.

PROPAGATE Divide in spring, cutting up rhizomes into sections in such a way that each section has its own growth buds at the top.

WAYS TO GROW

Arum and calla lilies are grown as summer bedding, as border plants, in pots, and for cut flowers outdoors, as well as indoor plants. Hardy types are excellent for bog or damp gardens or as marginal plants by or in a pond.

SEASONAL SCHEMES LATE SPRING TO MIDSUMMER

Gardeners often call the period from late spring to midsummer "the **May gap**", but with its abundance of terrific bulbs, this is one of the most exciting times of the year in the garden. The early spring bulbs may have faded out, but the next wave of blooming beauties are beginning to peak, and you can use them to create terrific displays in pots, borders, and beds. Partner up the right plants and you'll carry on the colourful show, and make magical mixes with herbaceous perennials and climbers.

SUMMER STUNNER

To create a sensational summer container scheme in one pot, the trick is to use three key plant types: a thriller, a filler, and a spiller. The "thriller" is often the tallest, most eye-catching plant, used to create height. "Filler" plants form round or mounding shapes in the middle of the planting, and bring schemes together by covering the spaces and bridging the heights between the thriller and the spiller. The "spillers" are often low-growing or trailing plants that tumble out or drape over the front and edges of the planter. Bulbs make terrific thrillers and fillers, and variegated ivy is a great evergreen spiller that looks good with a range of combinations through the year.

RECREATE IT A metal planter makes a fabulous focal point with a "thriller" butterfly gladioli (*Gladiolus* 'June') **(1)**, while purple *Penstemon* **(2)**, purple fountain grass (*Pennisetum advena* 'Rubrum') **(3)**, and dwarf Asiatic lily (*Lilium* 'Ladylike') **(4)** act as "fillers". Finally, the "spiller" here is pink ivy geranium (*Pelargonium peltatum*) **(5)**.

> **TOP TIP** IF YOUR CONTAINER WILL BE SEEN FROM ALL SIDES, PLANT A TALL "THRILLER" IN THE CENTRE. IF IT'S AGAINST A WALL OR ONLY SEEN FROM ONE SIDE, PLACE YOUR THRILLER AT THE BACK. FOR BEST RESULTS, CHOOSE ONE THRILLER, AND PLANT IT FIRST.

PRETTY PASTELS

At this time of year, gardens are blessed with an array of burgeoning seasonal blooms in cool pastel colours, which perfectly suit cottage-style garden schemes. These pale-toned flowers, in soft, soothing shades of light pink, lavender-purple, powder blue, primrose yellow, cream, and white are natural companions, and together create a relaxed, romantic feel. This is most successful where the bulbs and perennials of different forms and heights are intermingled, rather than planted in strict blocks or clumps.

RECREATE IT This jumble of perennials and summer-flowering bulbs includes sweet rocket (*Hesperis matronalis*) **(1)**, *Iris* 'Jane Phillips' **(2)**, *Rosa* ROYAL PHILHARMONIC **(3)**, foxgloves (*Digitalis purpurea* f. *albiflora*) **(4)**, bistort (*Persicaria bistorta*) **(5)**, foxtail lily (*Eremurus robustus*) **(6)**, and *Iris* 'Sparkling Rose' **(7)**.

TOP TO TOE

Flowering bulbs are the stars of borders during these months. To grow your own stunning borders, combine two or three different bulbs with a late-spring-blooming climber, such as wisteria, clematis, or laburnum, as a backdrop, and a low-mounding plant, such as lavender or lady's mantle, in front of the border to help camouflage their fading leaves.

RECREATE IT This border features pink spikes of Byzantine gladioli (*Gladiolus communis* subsp. *byzantinus*) **(1)**, blowsy white peony (*Paeonia lactiflora* 'Jan van Leeuwen') **(2)**, and purple globes of *Allium* 'Globemaster' **(3)**, with white Japanese wisteria (*Wisteria floribunda* 'Alba') behind **(4)** and African daisies (*Osteospermum* 'Stardust') in front **(5)**.

High summer brings bulbs in warmer shades, such as the honeyed blooms of red hot poker *Kniphofia* 'Tawny King'. Its upright flower spikes contrast well with the elegant arching stems of white angel's fishing rods.

MID- TO LATE SUMMER

This is the time of the year when the garden reaches its flowering peak: beds, borders, and pots burst forth with vibrant colours and layers of texture from a broad array of annual and perennial plants, but particularly from bulbs – the stars of summer. Enjoy the rich, warm tones of red hot pokers, or cool things down with white and blue African lilies. Add a touch of drama with large-leaved cannas and tropical-looking pineapple lilies, or contrast the upright spikes of blazing star with the graceful, arching stems of angel's fishing rods.

AFRICAN LILY *AGAPANTHUS*

Also known as lily of the Nile, these summer showstoppers feature large, domed flowerheads comprising loose clusters of small trumpet-shaped blooms. They come in a range of blues and purples or white, and sit atop tall, straight stems. Loved by bees, they also make great cut flowers.

BULB TYPE Rhizome
HARDY Fully hardy, half-hardy, tender
HEIGHT 0.4–1.5m (1⅓–5ft)
LEAF Strap-shaped; deciduous, evergreen
POSITION Full sun

CHOOSE

Some African lilies are deciduous and mostly hardy to -15°C (5°F), once established, while others are evergreen and more tender. Deciduous types include *Agapanthus* 'Margaret', one of the hardiest cultivars, with mid-blue blooms, which grows to around 90cm (36in). *A.* 'Midnight Dream' is also very hardy, with dark blue to purple-black flowers, to 70cm (28in). 'Brilliant Blue' reaches 60cm (24in), with lovely true-blue blooms. The best hardy, white-flowered variety is *A.* 'Arctic Star'. The popular Headbourne Hybrids are deciduous but borderline hardy.

Evergreen types include *Agapanthus africanus* and *Agapanthus praecox* and their cultivars. *A.* 'Queen Mum' is tall, to 1m (3ft), and bi-coloured, with white flowers with blue bottoms.

Two of the best compact growers are 'Dwarf Blue' and 'White Pixie', which both reach 40cm (16in).

PLANT AND CARE

African lilies need a position in full sun, in well-drained soil. Avoid wet, heavy soils – in these situations, grow in pots. Unlike most bulbs, which don't like to get congested, *Agapanthus* prefer to fill a pot, and flower better when roots are restricted – not pot-bound, but snug. Grow plants in a container that's about 5cm (2in) bigger than the plant itself, and pot up to the next size every few years as it increases.

Plant rhizomes in spring after all risk of frost has passed. Place about 5cm (2in) deep or just under the surface of the soil, with the pointed end facing up. Space at least 30cm (12in) apart.

Agapanthus **'Brilliant Blue'** is a compact, free-flowering variety.

Tall, evergreen A. 'Queen Mum' is two-tone white and blue.

WAYS TO GROW

Deciduous types look fantastic in borders, gravel gardens, and Mediterranean-style schemes. Evergreens are best in narrow south- or west-facing beds along a house wall in mild climates, and containers elsewhere.

Water plants in the ground well during the first season. Once established, they are quite drought-tolerant, but will flower best the next year if they have moisture in summer and early autumn. Container-plants need regular watering, and feeding with general liquid fertilizer from spring to early autumn.

Deadhead spent flowers, cutting out the whole stem. Remove foliage on deciduous types after it has died back on its own. Take old, damaged leaves off evergreen types when seen.

Evergreens should be grown in containers that are brought under cover or kept somewhere frost-free for winter. They're often grown in plastic pots that can be dropped into borders in late spring and lifted out in autumn.

Any *Agapanthus* grown in a container will need some protection over winter in colder climates, such as placing in a cold frame or wrapping in fleece. Mulch those left in the ground in autumn.

PROPAGATE Divide established clumps during the spring.

PERUVIAN LILY
ALSTROEMERIA

These showy clusters of funnel-shaped, lily-like flowers come in vibrant shades of red, orange, yellow, pink, and white, with striking markings, including stripes and splotches. The long-flowering blooms are produced from late spring to late autumn, with the main display in mid- to late summer.

BULB TYPE Tuber
HARDY Fully hardy, half-hardy
HEIGHT 0.15–1m (½–3ft)
LEAF Narrow, lance-shaped; deciduous. Spread 50cm (20in) or more.
POSITION Full sun, part shade
WARNING! Contact may irritate the skin. Wear gloves and other protective equipment. Wash hands after contact.

Alstroemeria INDIAN SUMMER (Summer Paradise Series) flowers for months.

A. ligtu has showy blooms that are great for cutting and last ages in a vase.

CHOOSE

Alstroemeria aurea is one of the taller species available, to 1m (3ft) high. Its cultivars 'Lutea', which is yellow with brown markings, and 'Orange King', which is bright orange with yellow patterning, can reach to 1.2m (4ft). Some of the most popular tall varieties are A. INDIAN SUMMER (Summer Paradise Series), with orange-yellow flowers, and the Planet Series cultivars, such as 'Saturne'.

A. *ligtu* hybrids are shorter, to 50cm (20in), and come in a range of colours, from cream to pink and red. The Princess Series cultivars are dwarf types. A. ROCK 'N' ROLL is something a bit different, with red blooms and variegated foliage.

PLANT AND CARE

Peruvian lilies prefer a sheltered spot in full sun or partial shade, and will grow in any moderately fertile, well-drained soil. They like some moisture while in growth but hate sitting in damp or wet conditions, especially while dormant.

Plant tubers in autumn, around 20cm (8in) deep, and spaced at least 30cm (12in) apart. Top with a deep mulch of chipped bark. If growing in containers, add grit to compost before planting.

Keep pot-grown plants well watered and fed in the growing season. Store pots in a frost-free place over winter.

There are Peruvian lily cultivars bred for the floristry trade that are less hardy and usually grown in greenhouses. However, once established, varieties sold as garden plants are usually frost-hardy, and those grown in the ground will withstand dips in temperature down to around -5°C to -10°C (23°F to 14°F). A. *ligtu* hybrids are known to handle down to -15°C (5°F). Add a dry mulch each autumn to protect them from frost. In colder climates, it might be best to lift and store the tubers somewhere frost-free over winter.

Taller flowers may need staking. Peruvian lilies might require protection from slugs in spring. Plants spread easily in the right conditions and are thought invasive in some warm climates.

PROPAGATE Divide established clumps by lifting and pulling apart tubers carefully in autumn or early spring.

WAYS TO GROW

These versatile plants look great in borders in both cottage-garden or bold, eye-catching schemes. Shorter and dwarf cultivars are the best for containers. They make fantastic cut flowers, but stems should be plucked or pulled rather than cut.

BEGONIA *BEGONIA*

Flamboyant, eye-catching, and long-blooming, begonias are the kings of summer bedding, bringing a touch of glamour to borders and containers with their exciting flowers and foliage. They'll flower all through summer until the first frosts with dazzling, abundant blooms in a diverse range of bright colours.

BULB TYPE Tuber
HARDY Tender, half-hardy
HEIGHT 15–50cm (6–20in)
LEAF Varied in size and shape; deciduous. Spreads to 50cm (20in)
POSITION Part shade, full sun

White-bloomed *Begonia grandis* subsp. *evansiana* var. *alba* has red leaf backs.

Nonstop Series plants have double flowers in vibrant colours.

CHOOSE

There are many different types of begonia, including the tender types that are available as houseplants, but the ones raised from tubers are some of the best for growing outside in the garden during summer. One of the most hardy is *Begonia grandis* subsp. *evansiana*, which can handle temperatures down to 0°C (32°F). Other popular cultivars include types such as the half-hardy 'Apricot Shades Improved', as well as the Nonstop Series, which, as the name suggests, are prolific bloomers, with double flowers in a mix of colours, including bright white, yellow, and red. For plants with a cascading or trailing habit for pots and baskets, look out for *B. boliviensis* hybrids such as the

Million Kisses Series, which have pointed green leaves and hanging, bell-like blooms in various shades.

PLANT AND CARE

Start off tubers under cover in early spring. Plant them at the surface of pots of compost, barely 2cm (1in) deep, with the hollow side facing up. Once all risk of frost has passed, harden the young plants off by placing outdoors during the day and bringing back inside at night for around a week, before planting them out in the ground.

Begonias need a sheltered spot out of the wind. They'll tolerate full sun but don't like to be too hot or dry, so do best in part shade. Plant out in fertile, well-drained soil. Water regularly, checking first by eye or with your finger. The soil should be moist but not wet. Avoid wetting the leaves, applying water at the base of the plant. Feed weekly with a fertilizer that's high in potassium, such as liquid tomato feed.

In autumn, once the foliage has yellowed, dig up the tubers and store over winter in a cool, frost-free place. Alternatively, lift and compost the plants and replace each year.

PROPAGATE Divide the tubers by cutting into sections, each with its own growth "eye" or bud, if growing under cover in spring, once leaves appear.

WAYS TO GROW

Begonias can be used to fill gaps in beds and borders, and create eye-catching displays in pots and hanging baskets.

The Million Kisses Series has pendent blooms and a trailing habit.

CANNA LILY *CANNA*

Canna lilies are big, bold, tropical-feel plants with large showy leaves, often in dark shades and variegated, and brilliant blooms in red, pink, orange, cream, and yellow. These tender but easy-to-grow perennials bring long-lasting colour, height, and drama to sultry summer schemes.

BULB TYPE Rhizome
HARDY Tender
HEIGHT Up to 2.5m (8ft)
LEAF Ovate, paddle-shaped; deciduous. Spreads to 50cm (20in)
POSITION Full sun, part shade

Add a touch of drama with the fiery flowers of *Canna indica*.

CHOOSE

Also known as Indian shot and arrowroot, *Canna indica* reaches up to 2.5m (8ft) tall, with banana-leaf-like foliage and upright stems topped with blazing red flowers. The cultivar 'Purpurea' has bronzed, purple-red leaves and orange flowers. TROPICANNA GOLD has variegated green and yellow striped foliage and flouncy, golden, iris-shaped blooms.

Canna 'Phasion' is a popular variety, growing to around 1.5m (5ft) tall, with orange flowers and pink-striped purple leaves. 'Striata' also has orange flowers, but reaches up to 2m (6½ft). 'Durban' is a reliable grower recommended for smaller gardens as it only reaches around 90cm (36in), with flamboyant,

multi-coloured leaves striped green, red, and pink with a sunset-orange flush and large orange flowers.

PLANT AND CARE

Canna lilies aren't hardy, so plant rhizomes under cover first in spring, sitting just under the surface in pots of compost. Give a little water, and grow on in a warm greenhouse, conservatory, or windowsill. Increase watering as shoots and leaves appear. When risk of frost has passed, place plants outdoors in the day and inside at night for about 10 days to condition them to outside temperatures. Then plant them out in the ground, in a sheltered spot in full sun or partial shade – those with dark leaves will colour up best in full sun, while those with green leaves and pale flowers will shine in partial shade. They need a rich, moist but well-drained soil. Plant at about 10cm (4in) deep, at least 75cm (30in) apart. You can also grow in large pots over 30cm (12in) wide.

Keep moist while growing, watering well if dry. Either feed weekly with high potassium, liquid tomato fertilizer from planting to when they start to flower, or regularly with a general-purpose fertilizer while in growth. Deadhead spent blooms to encourage more.

In mild climates and areas with little frost, like urban and coastal gardens, you could brave leaving canna lilies outside in the ground all year, but this is a gamble. Leave the foliage to collapse in autumn before folding over and

WAYS TO GROW

Canna lilies are used as summer bedding and excellent gap fillers, offering height at the centre or the back of borders. They're great in pots, and bring flair to hot-coloured, tropical or jungle schemes.

Canna **'Durban'** has striking, paddle-shaped, muti-coloured foliage.

topping with straw, or removing and covering with a deep mulch for protection. In colder areas, move pots somewhere frost-free or lift rhizomes once foliage has withered and store under cover over winter.

PROPAGATE Divide in spring before planting by cutting rhizomes into sections, each with its own growth'"eye" or bud.

ANGEL'S FISHING ROD

DIERAMA

These graceful perennials send out delicate arching stems that drip with pendent little bell-shaped flowers, and dangle like fishing rods above tufted hummocks of grassy foliage. They seem fussy, but give them the right growing conditions and the low-care lovelies will reward you year after year.

BULB TYPE Corm
HARDY Fully hardy
HEIGHT Up to 1.8m (6ft)
LEAF Grass-like; evergreen, semi-evergreen. Spreads to 60cm (24in)
POSITION Full sun

Dierama pulcherrimum has elegant arching stems and pink to purple blooms.

CHOOSE

Dierama pulcherrimum, also known as wandflower, is the most widely available species, and reaches 1.8m (6ft). It bears nodding blooms that resemble long, narrow funnels, in pink to purple shades – this particular species can be variable in form and flower colour. *D. pulcherrimum* var. *album* has blooms that are white or palest pink. *D. argyreum* is another white type, to just 1m (3ft) with delicate flower clusters that are pale yellow in bud and open to ivory. The hybrid 'Guinevere' is white with mauve centres, and grows to 1.4m (4½ft) tall.

Darker colours are also available, such as *D.* 'Blackbird', a popular dark purple-red cultivar which reaches 1.2m (4ft); and *D.* 'Blackberry Bells', with deep magenta, bell-like blooms. *D.* 'Miranda' has dark pink flowers, while compact *D. dracomontanum* comes in shades of coral pink.

PLANT AND CARE

Dierama are often believed to be susceptible to frost, or not fully hardy. However, most will survive temperatures down to around -10°C (14°F) in all but the most exposed gardens, as long as they're sheltered from cold, drying winds, and their roots aren't wet during winter. A little shade is fine, but they prefer a sunny spot in humus-rich, cool, well-drained soil, which stays moist in summer. Add grit and organic matter to the planting hole to improve drainage, and keep the roots cool under a gravel or pebble mulch.

Plant the corms in spring, around 7cm (2¾in) deep, and at least 50cm (20in) apart. Water in well and then continue to irrigate while the bulb establishes, and regularly during the growing season – daily in dry spells.

Dierama will act like an evergreen in milder climates and semi-evergreen in colder ones, losing some its leaves during the winter. Trim off old brown, tatty foliage in spring, when you can also boost with a feed of general-purpose fertilizer.

Give the plants some space as they don't like being crowded. Over time, they'll form large clumps, but they hate to be disturbed, so only divide if really congested. Expect them to "sulk" for a while afterwards, taking a few seasons to recover and flower again.

PROPAGATE Lift and divide clumps in spring, or after they've flowered, making sure to keep them moist. *Dierama* will self-seed prolifically, and you can collect the seed when ripe in late summer or early autumn and sow immediately, but the resulting plants may not be the same type as the original parent cultivar.

WAYS TO GROW

You can grow angel's fishing rods in borders and island beds but they'll really shine in schemes where they have space to show off, such as in gravel gardens, rock gardens, and displays with ornamental grasses. They're not suitable for containers.

PINEAPPLE LILY *EUCOMIS*

BULB TYPE Bulb
HARDY Half-hardy, tender
HEIGHT Up to 75cm (30in)
LEAF Lance-shaped; deciduous
POSITION Full sun

These extraordinary, cone-shaped blooms comprise masses of mini florets in white or pale purple, with a striking crown or bract of green leaves on top, reminiscent of the plant's namesake fruit. The exotic-looking flowers open from base to tip on short, stocky stems.

CHOOSE

Eucomis comosa is one of the most widely grown types of pineapple lily, with green-tinged white flowers to 75cm (30in) tall. The cultivar 'Sparkling Burgundy' is a real showstopper, with purple-hued pale blooms and dark reddish-purple stems and leaves.
E. bicolor has stunning white to green flowers with unusual purple-red lining on the petal edges and in the centres. It grows to around 50cm (20in) high.

Eucomis bicolor has a green crown of leaves and white, purple-edged petals.

E. comosa 'Sparkling Burgundy' has pale blooms with purple-hued leaves.

E. autumnalis has narrower, white flower spikes and plain green leaves, and reaches 40cm (16in).

PLANT AND CARE

Pineapple lilies are not hardy in colder climates, but can be grown outdoors all year in milder areas such as frost-free coastal and urban gardens, with a mulch in autumn for protection. Elsewhere, grow them in containers that can be brought indoors in winter, or lift the bulbs once the foliage has died back and store under cover.

They prefer a sheltered spot in full sun, in fertile, very well-drained soil. Start bulbs off in spring in pots of compost under cover and move or plant them outside only once all risk of frost has passed. Plant them around 15cm (6in) deep and spaced at least 30cm (12in) apart. Place extra grit in the bottom of the planting hole, and sit the bulb on the grit. If growing in containers, use a pot or bowl with shallow but broad dimensions. Use a loamy growing medium with grit for good drainage.

Keep pineapple lilies well watered throughout the growing season, as they need consistent moisture as well as sun to flower well. However, they must stay dry in winter or they'll rot.

PROPAGATE Remove offsets from bulbs in spring and grow on.

WAYS TO GROW

Pineapple lilies are eye-catching plants for containers and pots displayed on the patio or terrace, or down a flight of steps. They will bring drama and glamour to the centre of a border, perhaps as part of a tropical-feel scheme.

Pineapple lilies create drama and a welcome vertical accent in a border.

FREESIA *FREESIA*

Coveted around the world for their long-lasting, sweet-scented flowers, freesias make a real impact in the late summer garden. Each stem bears multiple trumpet-shaped, upward-facing blooms along the top side, in clear, bright colours from white and yellow to red, pink, and violet.

BULB TYPE Corm
HARDY Tender
HEIGHT 30–60cm (12–24in)
LEAF Narrow, lance-shaped; deciduous
POSITION Full sun, part shade

CHOOSE

The most widely grown types of freesia are *Freesia* x *kewensis* hybrids, which are often sold simply as an unnamed mix of different colours, or with group or series names, such as *Freesia* Rainbow mixture. Most are single blooms, but there are also double-flowered types.

PLANT AND CARE

Freesias have a reputation for not flowering reliably, and sending up only leaves, but the solution is simple – seek out "prepared" corms in spring.

These have been heat-treated to mimic the plant's preferred natural climatic conditions, which encourages them to flower well a few months after planting. You can start off the corms indoors in early spring and bring them on under glass if you have a greenhouse, but the most straightforward way to grow them is to plant them outside in pots or in the ground in mid- to late spring, once the temperature has risen above 5°C (41°F). In cold areas prone to late frosts, plant outside in late spring once all risk of frost has passed. Prepared corms will flower about 10–12 weeks after planting. For flowers over a long period

WAYS TO GROW

Plant freesias in groups in the border for impact, or in blocks in the cutting garden. They also work brilliantly in containers around seating areas, where their pleasing perfume and pure colour tones can be easily enjoyed.

from summer into autumn, plant a batch of corms every week from mid-spring to midsummer.

Choose a sheltered spot in full sun or in partial shade with several hours of sunlight a day, in fertile, well-drained soil. Plant corms about 5–8cm (2–3in) deep, with the pointed end facing up, spaced at least 8cm (3in) apart.

Flower stems may need staking. If you wish, feed with liquid tomato fertilizer while growing. Deadhead flowers to encourage more blooms.

Once flowering has finished, you can lift the plant and discard it on the compost, if treating as an annual; or allow the foliage to yellow and die back before lifting to store over winter. Since any corms you keep to replant the following spring won't be heat-treated, they may not flower or might take much longer to bloom. To guarantee success, many gardeners buy new, prepared corms every year.

PROPAGATE Offsets can be removed from corms when lifted in autumn.

Freesias are popular cut flowers for scented bouquets and arrangements.

SUMMER HYACINTH *GALTONIA*

BULB TYPE Bulb
HARDY Fully hardy, half-hardy
HEIGHT Up to 1.2m (4ft)
LEAF Upright, strap-shaped; deciduous
POSITION Full sun

These impressive scented blooms are borne on large flower spikes in shades of white and pale green. Each bulb produces tall stems topped with dazzling flowers comprising up to 30 small, waxy, bell-like florets, which dangle daintily over the eye-catching upright green foliage.

Stately *Galtonia candicans* impresses with its towering spikes of white blooms.

G. viridiflora has masses of long-lasting, scented, green-flushed flowers.

Summer hyacinth needs to stay moist during the summer, so water regularly while it's in active growth, especially during dry spells. However, it hates to be wet while dormant, so keep dry in winter. It can tolerate temperatures down to around -10°C (14°F) in all but very exposed gardens, so in mild areas, you could risk leaving bulbs in the ground over the winter, with a mulch for protection. In colder climates or gardens with heavy or wet soil, lift the bulbs once the foliage has died back, store somewhere frost-free undercover, and replant the following spring. Plants may need protection from slugs.

PROPAGATE Remove offsets from the bulbs in autumn and grow on in pots. When grown in the appropriate conditions, summer hyacinths will produce lots of seed, which can then be collected ripe and sown under cover straight away.

CHOOSE

Galtonia candicans, which can reach up to 1.2m (4ft), features many nodding white funnel-shaped blooms on a conical flowerhead, and looks a little like a tall, loosely arranged hyacinth – hence its common name of summer hyacinth. It also produces wide, grey-green, spear-shaped foliage.

Galtonia viridiflora sends up one or two tall sturdy stems per bulb, which grow to around 1m (3ft) high. The flowers are a pale green-washed shade of ivory-white.

PLANT AND CARE

For best results, don't plant bulbs outdoors until late spring as the blooms will be better quality and last longer if they flower from late summer into autumn than in the high-summer heat. Choose a spot in full sun, or in partial shade with several hours of sunlight a day. Summer hyacinths prefer humus-rich, fertile, moist but well-drained soil, so add some grit and organic matter to the planting hole to improve drainage, if necessary. Plant 8–10cm (3–4in) deep, at least 10cm (4in) apart. Water in well.

WAYS TO GROW

Grow summer hyacinths in pots and containers to place where you wish in the garden for a striking display, or plant them into borders to fill gaps left by spring bulbs such as annual tulips. They work well in gravel garden plantings with other late summer bulbs and perennials, in dramatic, eye-catching schemes.

SWORD LILY *GLADIOLUS MURIELAE*

BULB TYPE Corm
HARDY Half-hardy
HEIGHT 0.8–1m (2⅔–3ft)
LEAF Sword-shaped; deciduous
POSITION Full sun

The striking, subtly scented, bright-white blooms of sword lilies bring a welcome freshness to the garden in late summer, and keep the floral show going right into autumn. The pale, star-shaped flowers have a surprising dark-purple splash at their centre, and add luminosity to any scheme.

Star-shaped blooms of *Gladiolus murielae* are held on arching tubes.

CHOOSE

Gladiolus murielae, sword lily, is known by several other names, including *Acidanthera*, *Gladiolus callianthus*, and Abyssinian gladiolus. Unlike other gladioli, which have colourful spikes packed with masses of small blooms, sword lilies have just a few large, individual white flowers atop each stem, held aloft on arching long-tubed bases so that they almost dance in the breeze.

Sword lily flowers are lightly fragrant. The corms are planted and bloom later than other gladioli, extending the display well into autumn. These plants also have strikingly upright, sword-shaped green leaves, and reach up to 1m (3ft) tall.

PLANT AND CARE

Sword lilies need a sunny, sheltered position and well-drained soil. Start the corms off in pots undercover in spring, and plant out once all risk of frost has passed. Alternatively, wait and then plant them directly outside in late spring. Place them 20cm (8in) deep, with the pointed end facing upwards, and spaced at least 15cm (6in) apart. Water in well. Although they're relatively drought-tolerant, it's a good idea to water sword lilies regularly during dry spells in the growing season.

These plants are not fully hardy, but could be left in the ground over winter in mild areas. Many gardeners tend to treat them as annuals, simply lifting and discarding the plants after flowering has finished, and then replacing with fresh corms every year. If you'd like to keep the corms, wait until the foliage has died back, then lift and store them undercover somewhere frost-free, and replant in spring.

PROPAGATE Lift corms and remove offsets, which can be planted in pots and grown on.

WAYS TO GROW

Their elegant form makes sword lilies a perfect choice for contemporary urban garden schemes. Grow them in groups or drifts in the border with other plants, or on their own in pots and containers. They also make excellent cut flowers.

GLORY LILY *GLORIOSA SUPERBA*

Add a touch of drama to your summer garden with this showy climber, which has glossy green foliage tipped with clinging tendrils. The large, tropical-looking flowers, in mixes of red, yellow, pink, purple, red, and orange, have deeply reflexed petals with wavy edges, and prominent stamens.

BULB TYPE Rhizome
HARDY Tender
HEIGHT 1.2–3m (4–10ft)
LEAF Lance-shaped; deciduous
POSITION Full sun
WARNING! All parts are highly toxic. Ingestion may cause severe harm. Contact with rhizomes may irritate the skin. Wear gloves and wash hands after contact.

CHOOSE

Glory lily is also known as flame or fire lily and climbing lily. The most popular and widely available cultivar is 'Rothschildiana', which has spidery dark pink to red blooms with bright yellow-edged petals, and contrasting orange anthers on green stamens, which curve out below. 'Lutea' has green to yellow flowers, and 'Carsonii' features deep purple-red and yellow blooms. 'Sparkling Striped' is yellow with striped pink petals, and 'Tomas de Bruyne' is a good variety for cut flowers, in red and yellow.

Gloriosa **'Lutea'** has beautiful golden blooms with reflexed, wavy-edged petals.

PLANT AND CARE

Glory lily is tender and can be grown indoors all year, or outside as a fast-growing summer plant. It's best in a sheltered spot in full sun, in a large pot; as a climber, it needs support, such as an obelisk or trellis. The long, skinny, tuberous rhizomes are usually available in spring. If buying in store, choose rhizomes with small shoots emerging at one end, and handle carefully.

Glory lilies should be started off indoors. To plant, part fill the pot with a loam-based compost mixed with grit. Place canes in the pot so you can later tie in the growing plant and train it to its support. Plant several rhizomes to one pot, around 15cm (6in) apart. If shoots are emerging at one end, plant those rhizomes at a downwards, 45-degree angle, with the shoot end at the top facing upwards, and the tip around 5cm (2in) below the surface. If you can't see shoots, lay the rhizomes flat on their sides, on the soil about 5cm (2in) down. Fill up the pot with growing mix, to about 2cm (1in) from the top rim. Water in well. Place the pot in a frost-free, warm, bright spot. Keep soil moist once plants are in active growth: pot-grown glory lilies may need a daily water. Tie in shoots and stems gently as they grow. Place the pot outdoors after all risk of frost has passed.

You can keep the plant indoors over winter, or treat it as an annual, and lift and discard the rhizomes after blooming. To keep them, cut stems down as top

WAYS TO GROW

Glory lilies are an unusual and dramatic plant for pots in a heated conservatory or glasshouse, but many gardeners also grow them outdoors each summer as a showy seasonal climber, over a support in a container on the patio, or up a trellis on a warm wall.

Fiery G. *superba* 'Rothschildiana' looks terrific clambering up a support.

growth dies back in autumn, and bring inside. Stop watering until spring, when you can repeat the growth process.

PROPAGATE The rhizomes can be divided carefully in early spring – make sure that each section has a growth shoot or bud.

CORN LILY *IXIA*

These star-shaped flowers are borne on long wiry stems in tight clusters of up to 20 cheery star-shaped blooms. Each one has six petals in bright and brilliant colours, from white, pink, and purple to red, green, orange, or yellow, with contrasting darker centres.

BULB TYPE Corm
HARDY Half-hardy
HEIGHT Up to 50cm (20in)
LEAF Linear, sword-shaped; deciduous
POSITION Full sun

CHOOSE

Ixia viridiflora is widely available and has up to 12 striking pale blue-turquoise-washed flowers with dark purple-black centres per stem. *I. viridiflora* var. *minor* is slightly whiter in tone.

The cultivar *Ixia* 'Panorama' is variable, from pink-flushed white to hot pink flowers with darker centres, and prolific, with up to 20 blooms per stem. 'Hogarth' has cream blooms with a reddy-purple centre, and 'Marquette' has pale yellow petals and dark purple centres. 'Yellow Emperor' is a bright sunny yellow with a dark red base.

'Spotlight' is slightly shorter than the others, to about 40cm (16in), and has white flowers with deep pinky-red centres, and pink stripes lining the backs of the petals.

PLANT AND CARE

In their natural, warm-climate habitat, corn lilies grow over winter and flower in spring, but in colder regions, they're grown by gardeners for summer flowers from corms planted in spring. They won't survive cold, wet winters in the ground, preferring it mild and dry, and are often treated as annuals.

WAYS TO GROW

Corn lilies bring a touch of the glamour and splashes of colour to pots and containers, and fit in well in Mediterranean-style plantings and hot, sunny urban gardens. Use them in groups as filler flowers in the mid- to late summer borders.

Plant corms in spring, once all risk of frost has passed and the temperature is above 5°C (41°F). They need a sheltered spot in full sun with fertile, well-drained sandy or loamy soil. Place the corms around 10cm (4in) deep, spaced 8–10cm (3–4in) apart, with the pointed end facing up. Corn lilies can also be grown in pots that are sunk into the border to fill gaps and spaces left by lifted spring bulbs, and then lifted in the autumn. Water well after planting, and once in growth, keep the soil just moist.

In very mild, frost-free areas, you can risk leaving corn lily corms in the ground over winter, with a mulch for protection. Otherwise, lift them in autumn once the foliage has died back, dry them and store in a cool, frost-free place over winter. You can also simply lift and discard them after flowering.

PROPAGATE Remove offsets when lifting in autumn. In mild climates with the right conditions, they're known to self-seed.

Ixia viridiflora **var. minor** produces clusters of purple-centred white flowers.

The cultivar 'Yellow Emperor' has star-shaped yellow blooms with red bases.

RED HOT POKER *KNIPHOFIA*

Red hot pokers are distinctive plants with tall, bold flower spikes in vibrant shades, from red and orange to green, bronze-brown, and coral-pink. Long-lived, vigorous, and prolific flowerers, these perennials are attractive to pollinators and produce their fiery, torch-like blooms over a long season.

BULB TYPE Rhizome
HARDY Fully hardy, half-hardy
HEIGHT Up to 1.5m (5ft)
LEAF Strap-shaped; evergreen, deciduous. Spread 60–90cm (24–36in)
POSITION Full sun, part shade

Kniphofia **'Percy's Pride'** has densely packed yellow-green flower spikes.

CHOOSE

The many species and cultivars of red hot poker, also known as torch lily, come in a broad range of colours, heights, and flowering times. Some are hardier than others, but for exposed gardens in cold climates, a good choice is *Kniphofia caulescens*, with orange-pink to pale yellow flowers; it reaches to around 90cm (36in). Hardy cultivars include *K.* 'Green Jade', with pale green to white blooms, and 'Percy's Pride', with yellow-green flowers, both of which bloom from midsummer, and reach around 1.2m (4ft). 'Fiery Fred' is another top performer, with mid-orange and yellow blooms from midsummer through to early autumn. There are several interesting varieties in biscuit shades, such as 'Barton Fever',

Orange and yellow *Kniphofia rooperi* starts blooming in late summer.

with dark cream blooms, and 'Tawny King', which has pale apricot to creamy-yellow flowers and dark bronze stems, to 1.2m (4ft). Some of the most compact types, at around 90cm (36in), are coral 'Timothy' and orange 'Gladness', which is also one of the earliest to flower, from early summer.

One of the latest red hot pokers to flower is *K. rooperi*, from late summer, with rounded flowers in bright reddy-orange and yellow, to 1.2m (4ft).

PLANT AND CARE

Bare-root rhizomes may be available but red hot pokers are most often sold as plants in growth, in small pots.

They prefer a position in full sun in order to flower well, but they'll also tolerate some light shade, and need fertile, well-drained soil that's relatively moist during the growing season but isn't wet over winter.

Plant in spring, adding organic matter to the planting hole. Whether planting a rhizome or a plant, make sure the crown sits just at the level of the soil surface. Water in well.

Although they're quite drought-resistant, for best results, water red hot pokers regularly during dry spells. Remove any spent flower spikes to encourage more blooms. In late autumn, give a deep, dry mulch to protect the crown over the winter.

Leave the foliage alone until mid-spring, when you can tidy it up with shears or pull off tatty, brown, or dead leaves from the base. Watch out for damage from slugs and snails.

PROPAGATE Divide established clumps in the late spring.

WAYS TO GROW

Red hot pokers are great for coastal gardens, gravel gardens, and large rock gardens, as well as Mediterranean-style gardens. They tend to be too vigorous and grow too large for containers, but are excellent for adding vertical accents in planting, or at the back of a border.

BLAZING STAR *LIATRIS SPICATA*

Hardy and easy to grow, blazing star is a tough but pretty perennial that's low-maintenance and trouble-free, and a magnet for bees and butterflies. Its tall, striking flower spikes resemble fuzzy purple or white bottlebrushes, with long-lasting blooms that open in sequence from top to bottom.

BULB TYPE Corm
HARDY Fully hardy
HEIGHT Up to 90cm (36in)
LEAF Grass-like; deciduous
POSITION Full sun, part shade

Blazing stars are tough, hardy, and versatile plants for late-summer gardens.

Liatris spicata **'Alba'** sends up straight spires laced with white, feathery florets.

CHOOSE

Also called gayfeather, prairie star, and button snakewort, blazing star sends up poker-like, purple plumes above narrow, grass-like clumps of leaves. It reaches to around 70cm (28in).

The cultivar 'Alba' has spikes of fluffy white blooms. 'Kobold' is a more compact variety to just 50cm (20in), with lilac-pink flowers. The tallest forms, to about 90cm (36in), are 'Floristan Violett', which has magenta blooms, and white 'Floristan Weiss'. Both flower over a long period, until the first frosts.

PLANT AND CARE

Blazing star is tolerant of most growing situations. It prefers full sun but will tolerate some light shade, and will do well in any well-drained soil. The ideal conditions are some moisture in summer and dry winters, so for best results, add organic matter and grit to the planting hole to improve drainage. Plant corms in spring, straight into the ground, around 10cm (4in) deep and spaced at least 30cm (12in) apart, with the sprouted end facing up. Water in well.

Although it can cope with dry conditions, once established, blazing star should be watered regularly over its first few seasons of growth, to aid root development. Once sprouts appear, keep the soil lightly moist, until they die back in autumn.

The foliage and spent flower spikes can be cut down in the autumn once they've turned brown. Alternatively, you can leave them up, and cut them back in spring, so they provide structure over winter and also offer food to wildlife, such as birds, with their seedheads.

PROPAGATE Divide clumps and remove offsets in autumn, once foliage is dying back, or in spring as it emerges. Collect seed in late summer or autumn when ripe and sow immediately.

WAYS TO GROW

These blooms grow well in beds and borders, where they lend vertical structure, and in naturalistic and prairie-style plantings like perennial meadows. Compact forms such as 'Kobold' are great for containers.

Lilac-pink blooms of *Liatris spicata* 'Kobold' light up prairie-style schemes.

BUGLE LILY *WATSONIA*

Add some heat to the summer garden with these tall and tropical-looking blooms in shades of peach, orange, red, pink, and white. Bugle lilies have architectural, upright, sword-like leaves, and flower spikes comprising many small funnel-shaped florets that flare like trumpets towards the tips.

BULB TYPE Corm
HARDY Half-hardy
HEIGHT Up to 1.8m (6ft)
LEAF Lance-shaped; semi-evergreen
POSITION Full sun, part shade

Watsonia pillansii is a prolific bloomer with long-lasting bugle-shaped flowers.

CHOOSE

There are more than 50 species of *Watsonia*, and many cultivars and hybrids, but only a few of them are widely available for cultivation in the garden. One of the most popular of these is *W. pillansii*, Beatrice watsonia, which has magnificent, variable pale apricot to orange-red flowers on stems that reach up to 1m (3ft) tall.

Other good garden choices are *W.* 'Stanford Scarlet' which produces up to 12 bright red, tubular flowers per stem, and reaches 1.5m (5ft); *W. borbonica* 'Peach Glow', with dark peach blooms to 1.2m (4ft); and the Tresco hybrids such as *W.* 'Tresco Dwarf Pink', which has pink flowers, and is more compact at 60cm (24in).

PLANT AND CARE

Bugle lilies may tolerate temperatures as low as about -5°C (23°F), and in mild gardens in frost-free coastal regions and urban areas, they can survive outdoors over winter, with a protective mulch.

Short but sweet, *W.* 'Tresco Dwarf Pink' has pretty candy-pink blooms.

WAYS TO GROW

Bugle lilies are excellent plants for the greenhouse or conservatory, and for coastal gardens, as well as showy specimens for large containers and rock gardens. Plant in groups for impact in tropical-feel borders with canna lilies and pineapple lilies.

In colder climates, and those with wet winters, keep them under cover somewhere frost-free during winter.

Grow bugle lilies in a greenhouse or conservatory in a large display pot: move the pot outside in summer; bring it inside in autumn. You can also plant corms in the ground in spring once risk of frost has passed, and lift in autumn. Alternatively, grow in large plastic pots, which you sink into gaps in beds and borders, and then lift out in autumn.

Bugle lilies need a sheltered spot in full sun, and well-drained soil. Keep moist in summer, but dry in winter, and avoid waterlogging. Plant corms 10cm (4in) deep, spaced 30cm (12in) apart. Water in well. Once shoots appear, keep soil just moist, until flowering ends. If leaving in the ground, cut back flower spikes but leave the semi-evergreen foliage alone until spring, when any tatty leaves can be tidied up.

PROPAGATE Divide clumps every few years in autumn.

SEASONAL SCHEMES MID- TO LATE SUMMER

This is the hot, slow, and sultry season in the garden, when the days are longest, the sun is highest, and the borders have reached their prime. Gardening in these days is all about relaxing and enjoying the show, and doing small tasks such as deadheading flowering bulbs to encourage them to produce more blooms. You can also fill any gaps in beds and replant pots of faded spring bulbs with exciting tender and half-hardy summer bulbs. Keep everything well-watered in dry periods, to extend the display for as long as possible.

A TOUCH OF PARADISE

Bring a taste of the tropics to your garden with an exuberant scheme of tender and half-hardy summer blooms from bulbs, bedding, and perennials. Combine big leaves and eye-catching dark or variegated foliage with bold blooms in blazing hot shades of red and orange for maximum impact. Vibrant and stimulating, this style of planting brings together bulbs that were started off in spring under cover, such as dahlias and begonias, and spring-sown annuals such as castor oil plant, with bedding plants like busy lizzies. Add an extra dimension with perennial plants that are kept over winter in a greenhouse or conservatory and moved outside in summer, such as succulents like *Aeonium* 'Zwartkop', tender geraniums and fuchsias, and glory lily.

RECREATE IT This impressive container display in green and orange includes pineapple lilies (*Eucomis pallidiflora* subsp. *pole-evansii*) **(1)** and *E. autumnalis* **(2)** with variegated silver grass (*Miscanthus sinensis* 'Morning Light') **(3)** and striking, dark-leaved canna lilies (*Canna* 'Phaison') **(4)**. Below these sit pots of tender fuchsias (*Fucshia* 'Coralle') **(5)**, geraniums (*Pelargonium* 'Vancouver Centennial') **(6)**, and the tight rosettes of neat, compact echeverias (*Echeveria secunda* var. *glauca*) **(7)**.

ON A ROLL

Slopes, especially steep ones, can be difficult to manage in a garden because they're hard to access and more likely to be dry and have soil erosion. The best course of action is to plant them with hardy, fuss-free perennial flowers and foliage plants that are able to handle the conditions well, including mid- to late-summer bulbs and ornamental grasses. For best results, partner these with low, evergreen, drought-tolerant plants such as cotton lavender to complete a striking, long-lasting seasonal scheme.

RECREATE IT This bank is bulging with bulbs, such as African lilies (*Agapanthus* 'Midnight Star') **(1)** and A. 'Blue Rinse' **(2)** and angels' fishing rods (*Dierama* 'Guinevere') **(3)**, with blonde grass (*Stipa tenuissima*) **(4)** and day lilies (*Hemerocallis* 'Sammy Russell') **(5)**, as well as red hot pokers, montbretias, and lavender.

BEST BUDDIES

Always consider the conditions that bulbs need before combining them in a scheme – there's little point in planting a bulb that's drought-tolerant and likes it sunny alongside one that prefers it damp and shady. Planting bulbs that have similar requirements gives a far greater chance of success and makes maintenance considerably easier – you'll also often find that they naturally look good together.

RECREATE IT Summer hyacinth (*Galtonia candicans*) **(1)**, red hot pokers (*Kniphofia uvaria* 'Nobilis') **(2)**, and vervain (*Verbena bonariensis*) **(3)** make a fabulous mix of forms, textures, and colours, and all like full sun and moist but well-drained soil.

The captivating, coral-coloured flowers of *Dahlia* 'Classic Poème' are magnets for bees and butterflies, and continue blooming all the way through into autumn.

LATE SUMMER AND AUTUMN

As the long, hot days of summer begin to shorten and turn cool, you'd be forgiven for thinking that the garden will begin to quietly fade. Thankfully, however, a wealth of beguiling bulbs have bided their time, waiting for this exact moment to shine. There are utterly dazzling dahlias, with their cheery blooms in a multitude of colours and forms, and montbretias in sizzling-hot shades to add some punch to late-season planting schemes. There are small, delicate autumn crocuses, bright and beautiful coral drops and nerines, and starry schizostylis to prolong the floral show into late autumn; and unexpected surprises in the shade, with exquisitely intricate toad lilies and luminous lily turf.

AUTUMN SNOWFLAKE

ACIS AUTUMNALIS

BULB TYPE Bulb
HARDY Fully hardy
HEIGHT Up to 25cm (10in)
LEAF Linear, grass-like; deciduous
POSITION Full sun, part shade

Autumn snowflake – previously considered to be an autumn-flowering form of *Leucojum* (see p.72) – has slender red-brown or green stems topped with two or more nodding bell-shaped blooms. The flowers have pink-red bases and appear either before or with the tufts of thin, green, grassy foliage.

The pendent white bells of *Acis autumnalis* dangle off red stems.

CHOOSE

Acis autumnalis, autumn snowflake, carries up to four white florets on little stalks per stem, all flushed with pink to red at the base. It reaches 10–20cm (4–8in). The leaves come out after the flowers, and can keep growing until spring, when they begin a short dormant season.

Acis autumnalis var. *oporantha* and *Acis autumnalis* var. *pulchella* are both variable forms that may even be the same. Their flowers are conical and hang down on arched stalks, and appear at the same time as the leaves. They reach 15–25cm (6–10in). An interesting new form is *A. autumnalis* 'September Snow', which is almost pure white, and produces leaves at the same time as the flowers.

PLANT AND CARE

Experience varies, but autumn snowflakes are known to be hardy down to -15°C (5°F) in the right conditions, in a sheltered spot with well-drained soil. They may not be as hardy in very exposed, cold, or damp soils. Give them excellent drainage and a spot out of the wind for best results. They're recommended to grow in full sun, though they do well in light shade too. *Acis autumnalis* likes the soil moist but well-drained, while the other forms

and varieties prefer to stay mostly dry, with light watering during the growing season only when needed.

Plant the small bulbs in late summer or autumn, placing them 10cm (4in) deep and spaced around 8cm (3in) apart, or closer, but not touching. Water lightly after planting if the soil is dry.

Look out for damage from snails and slugs. The plants don't leave much mess but you can tidy up any remaining dead foliage in spring. Leave them alone to grow for several years before moving or propagating.

PROPAGATE When happy, autumn snowflakes clump up and multiply, and can be lifted, divided, and replanted into smaller clumps, and offsets removed, while in growth. They set seed freely, so collect ripe seed in autumn, and sow into small pots of compost. Grow on over winter in a cold frame or unheated greenhouse and pot on or plant out seedlings once they're large enough.

WAYS TO GROW

Autumn snowflakes are most at home in rock and gravel gardens, scree beds, and alpine schemes, such as in a trough with other low-growing plants. They also do well on the fringes of woodland schemes.

AMARINE × *AMARINE*

The true divas of the autumn garden, amarines are prolific bloomers with sensational, pink, lily-like flowers borne in clusters on tall, bare stems. Created by plant breeders originally for use as cut flowers, these perennials are very similar to nerines but larger, tougher, and showier.

BULB TYPE Bulb
HARDY Fully hardy, half-hardy
HEIGHT 60cm (24in)
LEAF Strap-shaped; semi-evergreen
POSITION Full sun
WARNING! Ingestion may cause stomach upset. Wear gloves and wash hands after contact.

The **mid-pink flowerheads** of × *Amarine tubergenii* 'Aphrodite' (Belladiva Series).

CHOOSE

Amarines are a cross between two very different plants: *Nerine bowdenii* and *Amaryllis belladonna*. These hybrids have the hardiness and strong growth of their parents, and the flowers are a mix between them in size and shape. Leaves are semi-evergreen, depending on the climate, glossy green, and strap-like, and may come after the flowers.

× *Amarine tubergenii* 'Zwanenburg' is a popular form with lightly scented, deep pink flowers, to around 60cm (24in). × *Amarine tubergenii* Belladiva Series are also widely available, with upright stems to 50cm (20in), topped by up to 10 trumpet-shaped blooms in shades of pink to almost white. 'Anastasia' bears rich, hot-pink flowers; 'Tomoka' is an intense magenta pink;

'Aphrodite' has pastel pink blooms with darker markings at the centre; and 'Emanuelle' is palest pink to white.

PLANT AND CARE

Amarines are hardy to between -10°C (14°F) to -15°C (5°F) as long as they're grown in a warm situation with good drainage. They'll grow in light shade, but flower best where they can bake in full sun and be dry in summer. They tolerate drought and poor soil, but hate to be wet or waterlogged. Ideally, give them a sheltered, sunny spot, such as at the base of a south-facing wall, and moderately fertile, well-drained soil.

Plant bulbs in spring, very shallowly, so the tips of the bulbs are just poking out of the soil. Space 15cm (6in) apart. Water when dry during the growing season, if necessary, but stop once flowering finishes and keep dry while dormant. You can feed with general fertilizer while in growth, if you wish. In cold climates, dress with a protective dry mulch over winter. Watch out for damage from slugs.

PROPAGATE Amarines do best when they're left to grow in dense, undisturbed, congested clumps, so allow them to establish, and only split up or propagate after several years, and when it becomes necessary for space or flowering vigour. Lift and divide them after flowering.

Amarines bring a burst of colour to autumn borders.

WAYS TO GROW

Amarines are versatile and will light up most positions in the garden, from borders to pots, and many types of garden, from cottage-style plantings and dramatic schemes to coastal sites and gravel gardens.

CORAL DROPS *BESSERA ELEGANS*

Bring some punchy colour to your seasonal planting with these perennial gems, which have upright, grass-like foliage and delicate clusters of fiery, reddy-orange flowers, like miniature ballerinas in tutus. The blooms dangle elegantly from stalks attached to the top of the long, slender stems.

BULB TYPE Corm
HARDY Tender
HEIGHT 60–70cm (24–28in)
LEAF Fine, narrow, linear; deciduous
POSITION Full sun, part shade

The red fuchsia-like flowers of *Bessera elegans* nod down elegantly.

directly in the ground in late spring. Alternatively, grow in pots indoors, with the containers brought outside in summer and back inside overwinter.

These plants prefer a light, sandy, or loam-based soil that's well-drained, and a position in full sun. Plant corms about 10cm (4in) deep, with the pointed end facing up, and spaced at least 10cm (4in) apart. Water in well. Wait until shoots appear, and then water regularly as needed during the growing season until flowering finishes.

If growing in the ground, leave foliage to die back on its own and then lift from the soil. Discard the old shrivelled corms and keep the new fresh ones. Store somewhere dry and frost-free over winter. In mild, frost-free gardens in coastal areas and urban centres, or warm climates, you could risk leaving corms in the ground overwinter with a deep, dry, protective mulch.

PROPAGATE Divide clumps and remove offsets while dormant.

CHOOSE

Bessera elegans is the most commonly grown and widely available form of coral drops. It has tall, bare, wiry stems up to 70cm (28in) high, which are crowned with dainty sprays of nodding flowers. The blooms have deep coral-coloured outer petals in a parasol shape, with light creamy stripes and patterns on the inside.

These pale colour markings match the cream central section that hangs down beneath.

PLANT AND CARE

Coral drops are tender plants and cannot grow outside all year in colder climates. Start off corms in pots indoors in spring and transplant out once all risk of frost has passed, or wait and plant

WAYS TO GROW

Coral drops make a real statement in summer borders and containers when planted in groups, especially in hot-coloured and tropical-looking schemes. They can be grown indoors in a glasshouse or conservatory, and are great long-lasting cut flowers too.

AUTUMN CROCUS

COLCHICUM AND CROCUS SPECIOSUS

Autumn crocuses are hardy perennials that offer a delightful surprise with their candy-coloured blooms appearing out of the soil, bare of foliage, in cheerful clusters. They're a wonderful foil to the changing leaves of trees and shrubs, and an excellent nectar source for late-flying pollinators.

BULB TYPE Corm
HARDY Fully hardy
HEIGHT 10–20cm (4–8in)
LEAF Lance-shaped or linear; deciduous
POSITION Full sun, part shade
WARNING! All parts of *Colchicum* are toxic. Do not ingest. Wear gloves and wash hands after contact.

Crocus speciosus **blooms** emerge on their own before the foliage.

CHOOSE

Two similar-looking plants are called autumn crocus: true *Crocus*, which flower in autumn; and those from a different genus called *Colchicum*, known as naked ladies or meadow saffron. *Crocus* are harmless, but *Colchicum* are toxic. You can tell them apart by their stigmas in the flower – *Crocus* have three, while *Colchicum* have six.

Autumn-flowering *Crocus* have the same cup-shaped blooms as spring crocuses. *Crocus speciosus*, with its lilac-purple blooms that appear in early autumn before the leaves, and its cultivars 'Conqueror' and 'Albus', are particularly popular. *Colchicum* flowers also emerge in autumn, on short bare stalks, in pink, purple, and white. The long, broad, strappy leaves usually appear in winter or spring, then die back for summer, when the plant is dormant.

Colchicum autumnale has mauve or white blooms. *C. speciosum* is taller, larger, and later-flowering. Good cultivars include 'The Giant', with large violet-pink flowers; and 'Waterlily', which has striking pink double blooms. *C. × agrippinum* has unusual snakeskin patterning on its pink petals.

PLANT AND CARE

Autumn-flowering *Crocus* prefers a position in full sun, in any well-drained soil, and needs to stay dry while dormant in summer. Plant the corms in

Colchicum **'Waterlily'** is an eye-catching double-flowered confection.

WAYS TO GROW

Autumn crocus look great in large groups, naturalized in short or fine grass, or planted under deciduous trees and shrubs. Plant in the middle of beds with other plants, rather than at the front, to help disguise the yellow foliage as it dies back.

late summer, around 10cm (4in) deep, spaced about 10cm (4in) apart, and water in well if there has not been rain.

Colchicum will grow in full sun or part shade, as long as it receives several hours of sun a day once its foliage emerges in spring. It will grow in drier conditions, but likes fertile, moist but well-drained soil. Plant corms in late summer, 8cm (3in) deep and 15cm (6in) apart, and water in.

Both types are low-maintenance once established. Water during active growth if conditions are hot and dry. Allow the leaves to die back on their own before removing – this also means that if they're planted in grass, you can't mow until midsummer. Watch out for damage from slugs.

PROPAGATE *Crocus* will self-seed and multiply through offsets. *Colchicum* can be lifted and divided or offsets removed in summer while dormant.

MONTBRETIA _CROCOSMIA_

Montbretias are vigorous, easy-to-grow, long-lasting flowers that will add a splash of hot colour to late summer borders in bright, fiery shades of red, orange, and yellow. The branching flowerheads comprise rows of small trumpet-shaped blooms, which line tall, arching stems above lush, spiky leaves.

BULB TYPE Corm
HARDY Fully hardy
HEIGHT 0.6–1.2m (2–4ft)
LEAF Sword-like; deciduous
POSITION Full sun, part shade

The cultivar **'Emberglow'** is neat but free-flowering with deep-red blooms.

CHOOSE

The most widely available montbretia is _Crocosmia × crocosmiiflora_, a variable hybrid, typically about 70cm (28in) high with orange flowers. One of the tallest, and earliest, cultivars to flower is 'Lucifer', which reaches 1.2m (4ft) and has striking, deep red blooms. 'Emberglow' is another good red form, with scarlet flowers with yellow centres, which reaches 75cm (30in). 'Carmin Brilliant' produces dark-pinky-red to orange flowers with paler centres, and reaches 50cm (20in).

Orange-flowered varieties include 'Emily McKenzie', 'Star of the East', and 'Severn Sunrise', which all grow to about 60cm (24in). Yellow-flowered types include 'Coleton Fishacre', 'Suzanna', and WALBERTON YELLOW.

PLANT AND CARE

Montbretia is a hardy plant and can tolerate light shade, but it will appreciate a sheltered position in full sun. It prefers humus-rich, moist but well-drained soil, and hates to be grown in wet, heavy conditions.

Plant in spring, after all risk of frost has passed. Place corms in the planting hole 8cm (3in) deep and at least 20cm (8in) apart, with the pointed end facing up. Water in well and keep moist during the growing season while they establish, and in dry periods. Those grown in pots will need to be checked and watered more regularly.

After flowering, leave the foliage to die back on its own. Mulch in autumn to offer them protection over winter.

Crocosmia × crocosmiflora **'Carmin Brilliant'** is a compact variety.

'Emily McKenzie' has orange flowers with deep red markings.

WAYS TO GROW

Montbretia works well in mixed borders, in late-season schemes with ornamental grasses, and in hot-colour and tropical-looking planting. Smaller forms look great in containers. It's also good for pollinator-friendly planting and as a cut flower.

These plants can become invasive given the right conditions and have a habit of jumping out of the garden and naturalizing in the wild. To prevent this from happening, plant them away from boundaries in country gardens, and don't put any plant material from them on the compost heap.

PROPAGATE Montbretia naturalizes happily if you let it – it multiplies easily and will need lifting and splitting every few years to ease congestion. Divide clumps and remove offsets in spring.

IVY-LEAVED CYCLAMEN

CYCLAMEN HEDERIFOLIUM

BULB TYPE Tuber
HARDY Fully hardy
HEIGHT 10–12cm (4–5in)
LEAF Rounded or angular, mottled; deciduous
POSITION Part shade

This compact perennial is a woodland wonder from late summer through to early winter, with slender stems carrying pink or white, occasionally scented, nodding blooms with deeply reflexed petals. They're followed by beautiful, ivy-shaped, dark-green leaves, marbled with silver markings.

CHOOSE

Also known as sowbread, ivy-leaved cyclamen has sweet, lightly fragranced flowers with gorgeous swept-back petals in shades of sugary pink, and later-showing silver-patterned arrow-shaped foliage. The cultivar 'Red Sky' has particularly deep pink to purple-red blooms.

Cyclamen hederifolium var. hederifolium f. albiflorum is the white version of the plant, and its cultivar 'Tilebarn Helena'

Cyclamen hederifolium has pale pink blooms and intricately patterned foliage.

is an exquisite, very hardy, and much sought-after, pure white form with pewter-flushed leaves.

C. hederifolium var. hederifolium f. hederifolium 'Ruby Glow' has magenta petals. 'Stargazer' is a really unusual form, with upward-facing flowers with pink tops, like little open mouths.

PLANT AND CARE

Ivy-leaved cyclamen will tolerate some sun, but are happiest in part shade in a sheltered spot. Grow in any well-drained soil and, for best results, add organic matter such as leaf mould to the planting hole. They like to stay dry while dormant in summer.

Plant tubers in late summer to early autumn, just under the surface of the soil, with the smooth side down and the side with the developing roots and shoots facing up. If you can't tell the difference, plant them on their sides – the tubers will right themselves in the ground. Give them space to grow: they will increase in girth, need good air flow around them, and mustn't be crowded by other plants. Water in well, and if it's hot and dry during the growing season, continue to water while they establish.

Mulch after the leaves have died back with a top dressing of leaf mould and grit. Once established, they're extremely low-maintenance plants and can be left undisturbed.

The white flowers and silver-washed leaves of 'Tilebarn Helena'.

PROPAGATE Ivy-leaved cyclamen don't produce offsets but will self-seed and spread happily and heartily by themselves. You can also collect seed when ripe and sow immediately.

WAYS TO GROW

Grow ivy-leaved cyclamen as groundcover under trees and shrubs, particularly in difficult dry shade areas, where they'll bring a punch of colour and pretty foliage from winter to spring. They also do well in rock gardens, alpine troughs, and containers.

DAHLIA *DAHLIA*

Dahlias are stalwarts of the late summer and autumn garden, producing an abundance of blooms right through to the first frosts. An enormous selection of plants is available – they vary in height, and flower form, size, and colour, with foliage ranging from green to darkest purple.

BULB TYPE Tuber
HARDY Tender, half-hardy
HEIGHT 0.6–2m (2–6½ft)
LEAF Divided; deciduous
POSITION Full sun

Deep pink **'Roxy'** is a Single-flowered, Group 1 dahlia.

'Totally Tangerine', from Anemone-flowered Group 2, has a bunched centre.

'Chimborazo' has the inner ring of petals of the Collerette types in Group 3.

CHOOSE

Most cultivars of herbaceous perennial dahlias grow to around 1.2–1.5m (4–5ft), though there are also dwarf varieties and taller types. The flowers also vary, from small 5cm (2in) diameter blooms to huge "dinnerplate" dahlias, 25cm (10in) across. They come in almost every colour, from red to white and yellow to purple. To help classify all the different types, they've been divided into several groups, based on the form of the flower.

GROUP 1 are Single-flowered, with simple daisy-like blooms, such as magenta *Dahlia* 'Roxy', yellow 'Bishop of York' and white 'Twyning's After Eight'.

GROUP 2 are Anemone-flowered, such as 'Totally Tangerine', which has pink outer petals, and an orange inner cluster of tubular petals. GROUP 3 are Collerette, like 'Chimborazo', which has an extra inner circle of smaller petals around the centre. GROUP 4 are Waterlily, such as shocking pink 'Kilburn Rose', with saucer-shaped double blooms just like a waterlily. GROUP 5 are Decorative, such as apricot-orange 'David Howard', dark red 'Arabian Night', and cream-pink 'Café au Lait' – all large double blooms with broad, pointed petals. GROUP 6 are Ball dahlias like purple-red 'Downham Royal', with geometric-looking, globe-shaped flowers. GROUP 7 are Pompon, similar to Ball types, but

smaller and perfectly round, miniature, honeycomb-textured spheres, such as lilac-pink 'Franz Kafka'. GROUP 8 are Cactus types, such as 'Hillcrest Royal', with masses of narrow, pointed petals; and GROUP 9 are Semi-cactus, with the same spiky look, but looser in form, as with bright red 'Indian Summer', and pale peach 'Preference'.

GROUP 10 is Miscellaneous, for dahlias that don't fit in other groups. It includes species dahlias, fringed Fimbriated types, Star forms like 'Honka Fragile', Double Orchid-flowered dahlias, like stripy 'Pink Giraffe', and Peony-flowered ones such as 'Bishop of Llandaff', with its famous red blooms and almost black foliage.

Dahlia **'Kilburn Rose'** is a Group 4 Waterlily variety.

PLANT AND CARE

Start your tubers off under cover somewhere that's bright and frost-free in early spring in large plastic pots, and plant them out after all risk of frost has passed – either by transplanting them into the ground or plunging the pots into gaps in beds and borders. Alternatively, you can plant the tubers outside directly in mid- to late spring.

Dahlias need a sheltered spot in full sun, and a rich, well-drained soil, which is preferably slightly acidic, and

'Café au Lait' is a popular Group 5 Decorative dahlia.

Pompon dahlia 'Franz Kafka', Group 7, has mauve mini globe flowers.

light. On heavy soils, add organic matter such as garden compost or well-rotted manure to the planting hole.

Dig a hole large enough for the tuber to sit in comfortably, and place it in with the "eyes" or growth buds facing up, so that the top of the tuber sits just under the soil, about 3–5cm (1¼–2in) down. Space tubers at least 60cm (24in) apart.

Dahlias often need support. Place stakes in the planting hole at the same time as the tubers – use posts for tall or large-flowered types and bamboo canes for smaller forms, placed around the plant. Tie them in as they grow.

Cactus dalias, like 'Hillcrest Royal', are in Group 8.

Water in well and regularly during the growing season, especially during hot, dry spells and those in pots. Once in growth, feed with liquid tomato fertilizer every two weeks. Deadhead flowers as they fade to encourage more blooms. Watch out for slug damage.

In mild climates with well-drained soil, you can risk leaving tubers in the ground all year, with a deep, dry mulch. In colder, wetter climates and heavy soil, lift the tubers in autumn once foliage has begun to die back. Store over winter in a frost-free spot (see p.22).

PROPAGATE Divide tubers in spring (see p.26).

Star dahlias, like 'Honka Fragile', are in the Miscellaneous Group 10.

WAYS TO GROW

Dahlias are great for any style of garden but especially for giving a boost to borders or beds with other late-flowering perennials and ornamental grasses, and hot-coloured schemes. Shorter cultivars are good in pots, and they make excellent cut flowers.

SCHIZOSTYLIS

HESPERANTHA COCCINEA

Seemingly delicate but truly tough, these late-flowering lovelies have pointed upright leaves below long bare stems. The stems are tipped with groups of star-shaped blooms in shades from white to coral-pink to dark red, from the end of summer through to the first frosts, and sometimes beyond.

BULB TYPE Rhizome
HARDY Fully hardy, half-hardy
HEIGHT Up to 60cm (24in)
LEAF Sword-like; semi-evergreen
POSITION Full sun

Hesperantha coccinea **'Major'** is a glorious, tough, floriferous variety.

CHOOSE

Also known as crimson flag, *Hesperantha coccinea* was previously named as *Schizostylis* and is still widely known and sold as such. *H. coccinea* 'Major' is the most popular cultivar – it's more robust than the species, with up to 10 deep red blooms per flower spike, and many stems per plant.

'Alba' is a white-flowered type with green centres. Palest pink 'Wilfred H. Bryant' is also sold as 'Pink Princess', while 'Mrs Hegarty' is light pink, striped with darker pink. Another good performer is 'Sunrise', which has large blooms in a peachy rose-pink.

PLANT AND CARE

Schizostylis are hardy enough to withstand -10°C (14°F), so they're usually grown outdoors all year as a perennial. However, for best results, grow them in a sheltered spot in full sun – such as at the base of a warm, south-facing wall – where they're moist in summer but not waterlogged in winter. Grow in any moderately

H. coccinea **'Sunrise'** has large pink flowers over a long season.

WAYS TO GROW

Schizostylis like moist conditions and as such are great to use around pond areas, but they'll also add welcome colour to late-season schemes in borders and in containers, and make excellent cut flowers. Pots brought into a greenhouse in winter may continue to flower all season.

fertile, moist but well-drained soil, and add organic matter to the planting hole to improve fertility and drainage.

Plant rhizomes in spring, around 10cm (4in) deep and at least 15cm (6in) apart. Water in well and keep well-watered during the growing season. Deadhead spent flowerheads to encourage more blooms. They'll bloom for longer in frost-free areas, sometimes right into winter. Cut down the flower stems once flowering has finished. The leaves are semi-evergreen, which means they may die back in cold winters, before fresh foliage appears the following season. Tidy up any brown or tatty foliage in late winter or early spring, if necessary.

Protect the roots of schizostylis by applying a deep mulch over winter.

PROPAGATE They'll spread when happy, bulking up quickly and becoming too congested to flower – although they're less vigorous on dry soils. Lift and divide in spring every few years.

LILY TURF
LIRIOPE MUSCARI

This unusual evergreen perennial produces low-spreading mounds of lush, grass-like foliage, and upright flower spikes of densely packed purple blooms, followed by black berries. Tough and easy to grow, lily turf looks great all year, and is drought-tolerant, so perfect for difficult dry and shady areas.

BULB TYPE Tuber
HARDY Fully hardy
HEIGHT 30–45cm (12–18in)
LEAF Linear; evergreen
POSITION Part shade, full shade

CHOOSE

Lily turf is easily mistaken for a grass because of its mounds of strappy leaves, but then surprises in late summer and autumn with short flower spikes to about 30cm (12in), covered in a tight cluster of vibrant violet-purple buds, which open to tiny star-shaped blooms. The dark berry-like seedpods come later.

Lily turf flower spikes feature masses of tightly packed, tiny, round, purple buds.

Liriope **'Monroe White'** will brighten up areas of deep shade.

The cultivar 'Big Blue' is taller than the species, to 45cm (18in), with bigger flower spikes, and grows best in part shade. 'Royal Purple' has darker purple flowers and can tolerate more sun than the other types. It's also a good plant choice if you want to attract late-flying pollinators. 'Monroe White' has large white flowers and prefers a position in full shade.

'Variegata' has larger leaves, striped with yellow or cream, but the most interesting foliage appears on 'Okina', with new spring foliage that's such a pale light-green it's almost cream or white; it gradually changes over summer until it becomes dark green in autumn.

PLANT AND CARE

Lily turf prefers fertile, acid to neutral, moist, well-drained soil, but is not fussy, and as long as it isn't waterlogged, it can tolerate a range of conditions, including drought. For best results, choose a sheltered spot in part shade.

Plant out from spring to late summer. They're usually available as potted plants; rarely as bare-root tubers. Plant pot plants so the soil is at the same level as it was in the pot. Plant tubers so the growing shoots or buds are sitting just under the soil surface. Water well and regularly in the first growing season.

Rake through the foliage with your hands in late winter to clear unsightly build-up. Remove old brown leaves and cut down old flower spikes. On new spring growth, look out for damage from slugs. Rabbits won't eat this plant.

PROPAGATE Lift and divide congested clumps in spring.

WAYS TO GROW

Lily turf is an excellent low-maintenance groundcover that will spread to block out weeds, and is often used en masse on slopes to prevent erosion. It's also highly valued for growing in shade, and under trees and shrubs, such as woodland schemes, as well as an edging plant for the front of borders.

NERINE *NERINE*

Striking and showy, nerines have dazzling pink, red, and white blooms atop long, upright stems. The strappy green leaves may emerge with or after the long-lasting, lily-like flowers, which comprise six recurved petals with ruffled edges, and are sometimes scented. Nerines are attractive to pollinators.

BULB TYPE Bulb
HARDY Fully hardy, half-hardy, tender
HEIGHT 40–60cm (16–24in)
LEAF Strap-shaped; deciduous
POSITION Full sun
SPREAD Ingestion may cause stomach upset. Wear gloves and wash hands after contact.

Nerine bowdenii offers clusters of stunning pink flowers with rolled-back petals.

CHOOSE

Nerine bowdenii, Bowden lily, is the only species hardy enough to grow out in the garden all year round in cool climates. It has mid-pink blooms and reaches around 40cm (16in) high, but its cultivars tend to grow to at least 50cm (20in) and come in a variety of reddish-pink to white hues. 'Isabel' has deep pink flowers, 'Nikita' has light pink, and 'Stefanie' the palest pink flowers. 'Alba' is pure white.

Nerine undulata Flexuosa Group are much less hardy and flower later, in late autumn to winter, but are an option for frost-free, mild, coastal,

and city gardens. However, they can get damaged by slugs and winter weather, so they're often grown in pots that are brought indoors in winter.

PLANT AND CARE

N. bowdenii is hardy to around -10°C (14°F), but needs a sheltered position in full sun, such as a warm spot at the base of a south-facing wall. The bulbs need to bake in summer to flower well.

Plant in late summer to autumn in well-drained, preferably poor to moderately fertile soil – nerines hate heavy soil and wet conditions. Add grit to the planting hole to improve drainage

if necessary. In mild areas, place bulbs so the top is just at the surface, with the nose exposed. In colder areas, place them about 5cm (2in) deep. Space bulbs around 10cm (4in) apart. Water in well.

Cut back the flower stems after flowering has finished if you wish, but allow the foliage to yellow and die back on its own before removing.

Hardy types will benefit from a deep, dry mulch for protection over winter. Half-hardy types should be grown in pots and brought undercover in autumn.

Those growing in containers should be repotted with fresh compost every few years in summer.

PROPAGATE Nerines can be propagated by division or chipping (see p.26). Leave bulbs undisturbed in the ground until clumps become so congested that flowering stops, because they grow best when the roots are tightly packed. Lift and divide when necessary in spring or early summer. The smaller clumps may take a year to re-establish.

WAYS TO GROW

Plant hardy nerines at the front or centre of a sunny border or to fill a warm bed along a wall. They look great in containers and flower best when they're congested. Half-hardy and tender types can be grown in a greenhouse or conservatory.

AUTUMN DAFFODIL *STERNBERGIA*

BULB TYPE Bulb
HARDY Fully hardy
HEIGHT 10–20cm (4–8in)
LEAF Narrow strap-shaped; deciduous
POSITION Full sun

Autumn daffodils shine right through the season with their upright, sunny, deep golden-yellow flowers on bare stems. They have long, strappy leaves with a faint pale stripe, which appear at the same time as the goblet-shaped blooms. These are perennial bulbs that will develop into clumps over time.

CHOOSE

Sternbergia lutea, known as the autumn daffodil or the winter daffodil, actually looks nothing like a daffodil, apart from the bright yellow colour of its flowers. Its form more closely resembles a crocus. You can tell the difference between the plants by the stamens – autumn daffodils have six, whereas crocuses have three.

The solitary yellow blooms usually grow to around 15cm (6in) high. *S. lutea* Angustifolia Group is more floriferous, with shorter stems and slender, more cone-shaped flowers, and a faint pale stripe on the leaves.

S. sicula 'Arcadian Sun' is smaller, to just 10cm (4in) or sometimes less, and often produces two blooms per stalk.

PLANT AND CARE

Autumn daffodils are hardy to around -10°C (14°F), but should be given some protection by planting in a sheltered spot at the base of a south-facing wall or fence, in full sun, and growing in a well-drained soil that's poor or gritty to moderately fertile. They won't do as well in heavy, wet soils.

Plant bulbs in late summer, placing them around 15cm (6in) deep, spaced around 10cm (4in) apart. Water in

WAYS TO GROW

Naturalize autumn daffodils en masse in a lawn or in a sunny spot under trees, or grow them at the front of a border. They do very well in pots outdoors or in a glasshouse, and are always a good choice for rock and gravel gardens.

Rockeries have the well-drained, gritty soil that autumn daffodils prefer.

Sternbergia lutea are crocus-like, with daffodil-yellow blooms.

well and then keep them just moist when in growth. They need to stay dry while dormant in summer.

Autumn daffodils can be affected by bulb flies, eelworms, and viruses, and fresh growth can be damaged by snails and slugs. They also hate to be disturbed, so can be left alone until flowering diminishes due to congestion.

PROPAGATE Divide clumps only when necessary for best results, every five years or so, when dormant in summer. Remove offsets at the same time.

TOAD LILY *TRICYRTIS*

These easy-to-grow, eye-catching woodland beauties have tall, upright, or arching stems and dainty flowers with petals spotted in purple, red, or brown. The orchid-like blooms are borne on their own or in clusters and are great for wildlife, as are the fruit-like seedheads that follow.

BULB TYPE Rhizome
HARDY Fully hardy
HEIGHT Up to 1m (3ft)
LEAF Lance or heart-shaped; deciduous. Spreads up to 60cm (24in).
POSITION Full shade, part shade
WARNING! Toxic to pets. Ingestion may cause stomach upset.

Tricyrtis formosana has dainty upturned starry blooms spotted with pinky-red.

CHOOSE

Tricyrtis formosana, toad lily, is a hardy perennial that grows to around 80cm (32in). It has green lance-shaped leaves with dark-purple spots, which grow straight off the stem, without stalks. It bears clusters of upward-facing, star-shaped white flowers dotted with dark pink. Stolonifera Group is more vigorous, spreading through stolons (runners), and taller, with upright stems to around 90cm (36in). 'Dark Beauty' is a pretty cultivar, to around 50cm (20in), with purple-spotted blooms; and 'Empress' has white flowers with red dots and yellow centres.

T. hirta, the Japanese toad lily, has hairy stems and leaves, and its white, reddish-purple-spotted flowers are trumpet-shaped with reflexed petals. 'Albomarginata', which grows to around 75cm (30in), is an interesting variegated cultivar with green heart-shaped leaves that are trimmed with white.

T. hirta is a woodland wonder with intricately patterned blooms.

PLANT AND CARE

Toad lilies are fully hardy, despite their delicate appearance, but they do appreciate being sheltered from the wind, with a position in full shade or part shade. Grow in a fertile, humus-rich, moist but well-drained soil, which is preferably acid to neutral.

Plant in spring once risk of frost has passed. Place the bare-root rhizomes 7–10cm (3–4in) deep, and spaced at least 15cm (6in) apart. Water in well and keep moist while they establish.

When the flowers fade, cut down the whole stalk to the base, but leave the foliage to die back on its own. Apply a deep, dry mulch to protect the plant in winter in cold areas.

Look out for damage from slugs and snails. Toad lilies can also be affected by a virus, which mottles the flowers.

PROPAGATE Lift and divide in spring, making sure each cut piece of rhizome has its own growth bud at the top.

WAYS TO GROW

Toad lilies brighten up woodland schemes in late summer and autumn, spreading to form large groups. They also add colour to dark corners of the garden. Grow them near ponds and in shady borders with other moisture- and shade-loving plants.

SOCIETY GARLIC

TULBAGHIA VIOLACEA

BULB TYPE Rhizome
HARDY Fully hardy, half-hardy
HEIGHT 30–60cm (12–24in)
LEAF Grass-like; deciduous, semi-evergreen
POSITION Full sun

Society garlic is so-called because when eaten it leaves a less pungent aroma on the breath than actual garlic. It has grassy tufts of leaves and tall, bare stems topped with clusters of long-lasting, mauve-coloured blooms. The attractive, starry flowers are often night-scented.

Society garlic has lilac-pink, allium-like flowers atop clear, slender stalks.

CHOOSE

Tulbaghia violacea, known as society or sweet garlic, is a deciduous perennial that grows to around 50cm (20in). Each long stem is topped with a group of up to 20 pale purple-pink trumpet-like florets, making a flower head that looks like an African lily, but looser. It can begin blooming as early as the end of spring, but keeps going over a very long period into autumn. The flowers have a sweet, hyacinth-like scent, but the narrow, grey-green leaves and stems release the smell of onions when crushed. Both are edible.

The most widely sold cultivar is 'Silver Lace', also known as 'Variegata', which is semi-evergreen in warmer climates and has pinker flowers and green leaves with silver edges. It flowers from midsummer to late autumn, and reaches about 45cm (18in), depending on growing conditions.

T. 'Purple Eye' is another notable variety worth seeking out. It's a smaller plant, to just 40cm (16in), with semi-evergreen leaves and pink flowers with dark centres.

PLANT AND CARE

Society garlic varies in hardiness: *T. violacea* is hardy in sheltered areas, but cultivars are often half-hardy. *T. violacea* is widely sold as a pond or marginal plant, but cultivars are suggested for dry and gravel gardens. In general, society garlic grows well in full sun, in fertile, preferably light, well-drained soil, preferring to be moist while in growth and dry while dormant.

Plant in spring, digging a hole wide enough for the rhizome to sit in comfortably. Place it so the top is just under the surface of the soil. Space plants at least 15cm (6in) apart. Water in and keep moist in spring and summer while it's growing, but stop watering once flower buds appear. Cut out spent flower stalks at the base as they fade to encourage more blooms.

Half-hardy types can be grown in pots and brought under cover in winter, but *T. violacea* should tolerate winter outdoors in mild city and coastal gardens. In colder climates, it's often grown in pots that can be kept in a sheltered area for winter. Alternatively, plant in a warm position in the garden, such as at the base of a south-facing wall, and dig up a small clump after flowering in autumn to grow over winter in a pot somewhere sheltered, as insurance. Cut back the rest and cover with a deep, dry mulch to offer protection.

PROPAGATE Society garlic will spread when happy, forming clumps and then large groups. Divide during spring every few years.

WAYS TO GROW

Grow in containers and sunny beds and borders. It looks wonderful planted en masse in dry, gravel, or Mediterranean schemes, or in a rock garden, and is great for coastal sites. *T. violacea* is often sold as a marginal plant for growing around water.

SEASONAL SCHEMES

The summer may now be drawing to its inevitable close, but the delights just keep on coming, with plenty of bulbs starting to send up their bright blooms as the season flows into autumn. Shady schemes are refreshed with colourful carpets of woodlanders under the fiery-hued turning leaves of deciduous trees and shrubs; and in borders, beds, and containers, bulbs combine magnificently with perennials, late-flowering annuals, and grasses to extend the show up until the first frosts.

WOODLAND WAYS

Shade plantings and woodland-style schemes get a real second wind during the autumn with an array of excellent bulbs, including autumn crocuses and ivy-leaved cyclamen, brightening up the earth beneath trees and shrubs just as their earlier-flowering relatives did in spring. A successful seasonal scheme can be achieved by partnering these dainty blooms with interesting textures from plants that have strappy, grassy foliage, and groundcovers that can cleverly cover bare soil and around leafless flower stems.

RECREATE IT This colourful, low-growing combination alongside a path includes the striking foliage of evergreen perennial black mondo grass (*Ophiopogon planiscapus* 'Kokuryu') **(1)** offering the perfect backdrop for the bright pink giant meadow saffron (*Colchicum speciosum* 'Atrorubens') **(2)**. It's surrounded by bugle (*Ajuga reptans* 'Catlin's Giant') **(3)**, whose bronzed purple-green leaves complement the purple flowers of lily turf (*Liriope muscari*) **(4)** across the way.

TOP TIP PLANT AUTUMN-FLOWERING, SHADE-TOLERANT BULBS IN GROUPS OF THREE TO FIVE. APART FROM LOOKING GOOD, THESE GROUPINGS MAKE IT EASIER FOR LATE-FLYING POLLINATORS TO FIND THE FLOWERS.

DRAMATIC PINKS

The magnificent blooms of autumn-flowering bulbs such as nerines, amarines, and schizostylis cause a tremendous stir in the garden with their bright colours and fabulous forms. However, it can be a challenge to find good companions to set off their glamorous looks. For best results, focus on late-flowering perennials such as hardy asters and chrysanthemums or half-hardy autumn sages in similarly striking shades of pink, red, and purple. Add a soothing contrast with silver or grey-green foliage plants such as silver spurflower or wormwood.

RECREATE IT Brilliant-pink Bowden lily (*Nerine bowdenii*) **(1)** grows happily in a narrow bed along a warm, sunny wall, in a sweet-as-candy pairing with a pot of autumn sage (*Salvia greggii* 'Icing Sugar') **(2)**. The plush, silver foliage of a liquorice plant (*Helichrysum petiolare* 'Goring Silver') **(3)** acts as a fantastic foil to the flowers and brings a welcome cool note.

SPECTACULAR FINALES

Many herbaceous borders hit a lull now, when the summer flowers have finished blooming. You can, however, use bulbs — along with perennials such as sedums, which are long-flowering, or those, like hardy geraniums, that will reflower when cut back — to design a sensational second scheme that reaches its peak from late summer into autumn.

RECREATE IT Golden semi-cactus dahlias (*Dahlia* 'Glorie van Noordwijk') **(1)** work well here with tall, airy vervain (*Verbena bonariensis*) **(2)**, which bloom from midsummer on to late autumn. Orange-yellow montbretias (*Crocosmia* 'Zambesi') **(3)** repeat the warm and cool mix of colours with purple catmint (*Nepeta racemosa* 'Walker's Low') **(4)**.

INDEX

Bold text indicates a main entry for the subject.

Author Stephanie Mahon

AUTHOR ACKNOWLEDGMENTS

Thank you to Diana Loxley and the team at Cobalt for all their hard work, and John Campbell for his help and support.

PUBLISHER ACKNOWLEDGMENTS

DK would like to thank Mary-Clare Jerram for developing the original concept, Margaret McCormack for indexing, and Paul Reid, Marek Walisiewicz, and the Cobalt team for their hard work in putting this book together.

PICTURE CREDITS

The publisher would like to thank the following for their kind permission to reproduce their photographs:

Alamy Stock Photo: adrian davies 117bl; Avalon.red 75br; Barrie Sheerman 100c; BIOSPHOTO 56cl; blickwinkel 8cl, 20cl; Botany vision 52cl, 89cr; CHRIS BOSWORTH 37t; Chris Mattison 74bl; Christopher Burrows 118bc; Clare Gainey 85tl; Deborah Vernon 2c, 14c, 21cl, 21br; 52cr, 53cr; Don Mennig 60c; Elizabeth Debenham 10bl; Ernie Janes 8br, 11bl, 45tl, 87bl; Ester van Dam 70cl; flafabri 137bl; Florapix 90br; Garey Lennox 127br; Gary Cook 68bl; Glenn Harper 53cl; Graham Prentice 11br; Henk Vrieselaar 68bc; Hervé Lenain 69cl; Holden Wildlife 92cr; Holmes Garden Photos 75cl; imageBROKER 49br, 95tl; INTERFOTO 46c; Joel Douillet 44cl; John Glover 43bc; John Martin 93bc; John Richmond 54c, 63bl, 82c, 85tc, 138bc; Julie Davenport 56bc; Kaliantye 19br; Kay Ringwood 29tr; KTT 29br; LEE BEEL 57br, 105bl; lemanieh 20tr; Louis Berk 43tl; Malcolm McMillan 51br; Marc de Boer 9br; Margaret Welby 67br; Martin Hughes-Jones 115cl, 123br; Matthew Taylor 53bc; MBP-Plants 32cl; mediasculp 101bc; mike jarman 98bl; Miriam Heppell 92c; Nataliya Nazarova 114bl; Nigel Cattlin 28br, 29tl, 29bc; Organica 94cl; P Tomlins 127tl; Panther Media GmbH 72cl, 137cr; R Ann Kautzky 135c; RM Floral 42br, 87bc, 95cr; Ros Crosland 13tr; Selwyn 45br; Sergey Kalyakin 48cl; Steffen Hauser / botanikfoto 67tr, 79tr, 135bl; Stephen Power 109cr; The National Trust Photolibrary 34br; Tim Gainey 17bl, 30c, 54cl, 93br; Tim Slater 8tr; Verena Matthew 132c; Wieslaw Jarek 6c; WILDLIFE GmbH 99br; Wiskerke 79br; Yakoniva 17tl; Yon Marsh Natural History 21cr.

Cobalt ID: 26tr; 26cr.

Dorling Kindersley: Alan Buckingham 28bl; Brian North 110br; Mark Winwood / Alpine Garden Society 42cr, 66cr; Mark Winwood / Avon Bulbs 26bl, 52bl; Mark Winwood / Dr Mackenzie 53tl, 53tc, 55bl; Mark Winwood / Hampton Court Flower Show 2014 4cl, 52br, 97bc, 108bl; Mark Winwood / Lullingstone Castle, Kent 22br; Mark Winwood / RHS Chelsea Flower Show 2014 113tc, 133tl, 133cr, 133bl, 77bl; Mark Winwood / RHS Wisley 12bl, 13br, 16tr, 22c, 41bl, 41br, 48br, 49tc, 55cr, 57tl, 57tc, 62cl, 65bl, 65cr, 66cl, 76c, 78bl, 78bc, 84cl, 84br, 85bc, 85br, 91cl, 92cl, 96bl, 96bc, 99tl, 102cl, 102br, 106c, 108bc, 109cl, 111cl, 113cr, 118bl, 119cr, 124c, 128cl, 129bc, 130cl, 130bc, 131bl, 131cr; Neil Overy 126cl; Peter Anderson / National Dahlia Collection 133bc; RHS Tatton Park 103cr, 113bl.

GAP Photos: 32tr, 32cr, 34cr; Abigail Rex 33tl; Christa Brand - Weihenstephan Trial Garden 9tl; Clive Nichols - Ulting Wick, Essex 141tl; Friedrich Strauss 10cr, 25tr, 81br, 104bl; Howard Rice 18cl; J S Sira 40cl; J S Sira - Designer: Hay Young Hwang, Sponsors: LG Electronics 105tr; Jonathan Buckley 24tr, 24bl; Jonathan Buckley - Demonstrated by Carol Klein 25c, 25cr, 25bc, 25br; Jonathan Buckley - Design: Carol Klein 80cr; Jonathan Buckley - Design: Sue and Wol Staines 39br; Lee Avison - Design: Maureen Sawyer / Southlands12 122bl; Leigh Clapp 27br; Marcus Harpur 140bl; Nicola Stocken 10bc, 35tl, 41tl, 58cl, 59bl, 141br; Nicola Stocken - Designer: David Ward 40br; Paul Debois 39tl; Richard Bloom 33br, 59tc; Richard Bloom - Garden: Wildside Plants - Designer: Keith Wiley 123t; Richard Bloom - Rod and Jane Leeds garden, Suffolk 11tl; Robert Mabic 27c, 27cr, 27bc, 81t; Tommy Tonsberg 27tr.

Getty Images: Jeff Overs 17cr; Ron Evans 38cl; Jacky Parker Photography 88cl; Veena Nair 100cl.

Illustrations by Cobalt id.

All other images © Dorling Kindersley

Produced for DK by
COBALT ID
www.cobaltid.co.uk

Managing Editor Marek Walisiewicz
Editors Diana Loxley, George Arthurton
Managing Art Editor Paul Reid
Art Editor Darren Bland

DK LONDON

Project Editor Amy Slack
Managing Editor Ruth O'Rourke
Managing Art Editors Christine Keilty,
Marianne Markham
Production Editor Heather Blagden
Production Controller Stephanie McConnell
Jacket Designers Nicola Powling, Amy Cox
Jacket Co-ordinator Lucy Philpott
Art Director Maxine Pedliham
Publisher Katie Cowan

First published in Great Britain in 2022 by
Dorling Kindersley Limited
DK, One Embassy Gardens, 8 Viaduct Gardens,
London, SW11 7BW

The authorised representative in the EEA is
Dorling Kindersley Verlag GmbH.
Arnulfstr. 124, 80636 Munich, Germany

Copyright © 2022 Dorling Kindersley Limited
A Penguin Random House Company
10 9 8 7 6 5 4 3 2 1
001-326190-Jan/2022

A CIP catalogue record for this book
is available from the British Library.
ISBN: 978-0-2415-3050-4

Printed and bound in China

For the curious
www.dk.com